Channels, Carriers, and Pumps

DR JOHN T. BROSNAN
DEPT. OF BIOCHEMISTRY
MEMORIAL UNIVERSITY OF NEWFOUNDLAND
ST. JOHN'S, NF CANADA A1B 3X9
PHONE (709) 737-8540
FAX (709) 737-4000

Channels, Carriers, and Pumps
An Introduction to Membrane Transport

WILFRED D. STEIN
Department of Biological Chemistry
Institute of Life Sciences
The Hebrew University of Jerusalem
Jerusalem, Israel

ACADEMIC PRESS, INC.
Harcourt Brace Jovanovich, Publishers

San Diego   New York   Boston
London   Sydney   Tokyo   Toronto

Academic Press, Inc.
San Diego, California 92101

United Kingdom Edition published by
Academic Press Limited
24–28 Oval Road, London NW1 7DX

Library of Congress Cataloging-in-Publication Data

Stein, Wilfred D.
    Channels, carriers, and pumps : an introduction to membrane
transport / Wilfred D. Stein.
        p.   cm.
    Includes bibliographical references.
    ISBN   0-12-665045-4   (alk. paper)
    1. Biological transport, Active. 2. Ion channels. 3. Biological
transport. 4. Carrier proteins.   I. Title.
QH509.S73   1990
574.87'5--dc20   90-486
                                                                 CIP

Printed in the United States of America
90 91 92 93   9 8 7 6 5 4 3 2 1

To my friend and colleague
William Robert Lieb
for his major contributions
over many years
to our joint analyses of
transport and diffusion

Contents

Chapter 6. Primary Active Transport Systems

Chapter 7. The Regulation and Integration of Transport Systems

Appendix 1. Single and Triple-Letter Codes for the Amino Acids

Appendix 2. Fundamental Constants, Conversion Factors, and Some Useful Approximations

Appendix 3. Relationship between Permeability Coefficient P_s and the Half-Time $t_{1/2}$ of Entry of a Permeant

Index

Preface

It has been a wonderfully exciting experience, these past three decades, to take part in that great scientific adventure which has led to our present understanding of the channels, carriers, and pumps, the subject of this book. Many of these entities which, thirty years ago, had been mere hypotheses, models, or concepts, have now been isolated and shown to be proteins, and the genes coding for these proteins have been cloned and sequenced. In one or two cases, these transport proteins have been crystallized and their three-dimensional structures determined by crystallographic methods. We know a good deal about transport kinetics and about what regulates transport through the channels, carriers, and pumps. But we do not have, even in one instance, a clear understanding of how these molecules function. We do not know how they distinguish so effectively between their substrates and other similar ions or molecules. We do not know how the transported substrates move through the transporting proteins, nor how these transporting proteins catalyze those movements and, in some cases, link transport to the consumption of metabolic energy.

This book is meant as an introduction to these subjects so that the reader can learn what is already known about them and can prepare to take part in the exploration of the unknown. Readers of this book will thus include those who are about to begin working on membrane transport or have just begun to do so. In addition, those working, for instance, on membrane-bound receptors who want to understand how these receptors function will find this book useful as an introduction to the transport proteins which the receptors help to regulate. The beginning neurobiologist will read it to learn about the function of the channels, carriers, and pumps that are the basis of nervous behavior at the molecular level. The pharmacologist-to-be will find that it provides an approach to understanding cellular and intercellular behavior at the membrane level.

A major aim of the book is to furnish a link between the experimental basis of the subject and theoretical model-building. Many examples of experimental work are presented in the illustrations, many taken from original papers, others recalculated and redrawn. Scattered throughout the book are "boxes" in which material of special, often highly current, interest is developed in a detail that might otherwise interfere with the flow of thought in that particular chapter. These boxes are meant, however, to be studied carefully, and form an integral part of the whole. The selected reference lists at the ends of the chapters are not intended to be comprehensive, but should be sufficient to support the arguments presented and to provide the reader with an entry into subjects that hold a special attraction. My earlier book, "Transport and Diffusion across Cell Membranes," published by Academic Press in 1986, provides a more comprehensive bibliographic source.

It is a pleasure to acknowledge the help I have received from colleagues who read part or all of earlier drafts of this book. In particular, Hagai Ginsburg, Steven Karlish, and Chana Stein suggested many important improvements. I am much indebted also to an anonymous reader, brought to this task by Academic Press, whose penetrating criticisms, based on a depth of knowledge and understanding, suggested that he or she should really have been the person to write this book! In this and other matters, the staff of Academic Press has been most helpful. Finally, I am very grateful to those who have given permission for the reproduction in this book of their original illustrations, as acknowledged in the figure legends.

Wilfred D. Stein

List of Symbols

Symbol[a]	Definition	Page[b]
$\bar{V}$	Partial molal volume of water	59
V_{max}	Maximum velocity	80
Z	Valence	34
zt (superscript)	Zero trans	150
ε	Dielectric constant	74
π	Osmolar concentration	59
ψ	Electrical potential	34
σ	Reflection coefficient	64
00 (subscript)	Zero concentration at both faces	168
1 (subscript)	Pertaining to side 1 of membrane	142
12 (subscript)	In direction side 1 to side 2	167
2 (subscript)	Pertaining to side 2 of membrane	142
21 (subscript)	In direction side 2 to side 1	167
I (subscript)	Pertaining to side I	30
I→II (subscript)	In direction side I to side II	31
II (subscript)	Pertaining to side II	30
II→I (subscript)	In direction side II to side I	39

[a] Symbols used in restricted context are defined locally and are not included in this list.
[b] Page on which first mentioned.

CHAPTER 1

Structural Basis of Movement across Cell Membranes

How do molecules and ions move across cell membranes? How does the cell membrane act as a barrier to such movements? How do special components of the membrane enable specific substrates to overcome this barrier and even allow metabolites to be concentrated within the cell or actively extruded from it? We shall try to answer these questions in this book. But first we need to look at the structure of cell membranes, since this structural information is essential in order to understand how all membrane transport takes place.

1.1. MEMBRANE STRUCTURE: ELECTRON MICROSCOPY OF BIOLOGICAL MEMBRANES

All living cells are enclosed by one or more membranes, which define the cell as a living unit—cutting it off from its environment. Figure 1.1(a) shows a cross section of the membrane that encloses the human red blood cell. (The picture was made using an electron microscope, the cell having been "fixed" prior to microscopy by treatment with a solution of potassium permanganate, which stabilizes the components of the membrane.) The cell membrane is seen as a double line of material lying diagonally across the photograph. The cytoplasm of the cell is the darker material lying to the right and above this line, while the extracellular environment lies below and to the left of the double line. The thickness of the membrane is some 7.5 nanometers (nm) [75 angstroms (Å)].

1

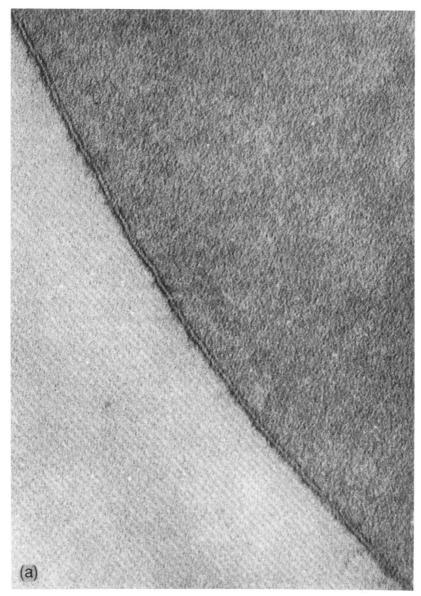

Fig. 1.1.   Structure of the cell membrane. (a) An electron micrograph of portion of a human red blood cell, fixed with permanganate and sectioned. Magnification: ×280,000. Taken, with kind permission, from J. D. Robertson, *in* "Cellular Membranes in Development" (M. Locke, ed.), pp. 1–81. Academic Press, New York, 1964. (b) Etched, freeze-fracture image of a ferritin-conjugated red cell membrane. Magnification: ×57,000. Taken, with kind permission, from D. Branton, *Philos. Trans. R. Soc. London, Ser. B.* **261,** 133–

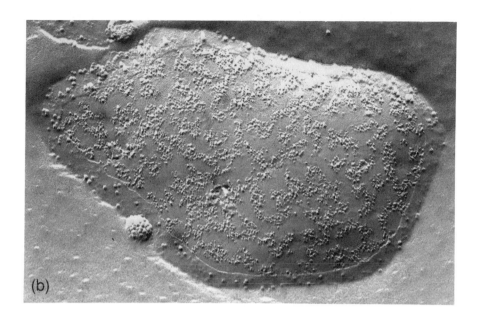

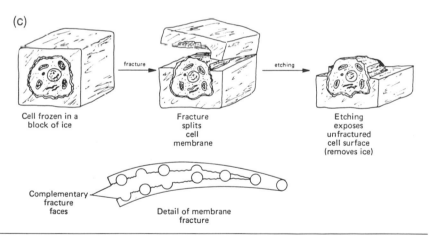

138. (c) Diagram to show freeze-fracturing and freeze-etching of a cell membrane. In "fracture" a glass knife is used to split the membrane between its hydrophobic layers. In "etching" the superficial ice is sublimed away to reveal part of the underlying surface. Taken, with kind permission, from R. D. Dyson, "Essentials of Cell Biology." Allyn & Bacon, Boston, Massachusetts, 1978. (d) Model of the fluid, amphiphilic membrane. The lipid molecules are depicted as the bilayer of oval heads with two tails, the protein molecules.as the large irregular shapes embedded in this bilayer. Taken, with kind permission, from B. Alberts *et al.,* "Molecular Biology of the Cell," 2nd ed. Garland Publishing, New York, 1989. (*Figure continues*)

(d)

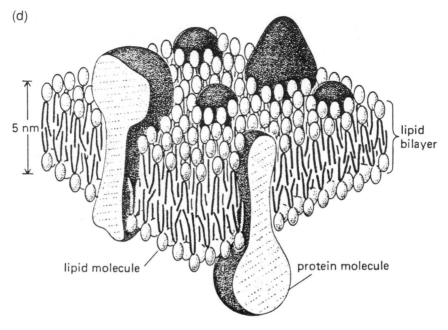

5 nm

lipid
bilayer

lipid molecule

protein molecule

Fig. 1.1.   (*Continued*)

Figure 1.1(b) is an electron micrograph looking down on the surface of the membrane of the human red blood cell. In this case, the membrane was prepared by the "freeze-fracture" method [Fig. 1.1(c)], which allows a researcher to look at the structure *within* the plane of the membrane. [Cell membranes are rapidly frozen and then fractured with a glass knife. This freeze-fracturing splits the membrane along a plane that appears to lie at the middle of the two lines seen in Fig. 1.1(a). The frozen and fractured specimen can then be "etched," a process in which the ice layers in a freeze-fractured specimen are sublimed away, exposing also the outermost surface of the membrane.] Look first at the convex surface in the center of Fig. 1.1(b). This shows the inner surface of half of the membrane, the half that lies on the right of the double line in Fig. 1.1(a). This is the face that is adjacent to the cytoplasm of the cell. We can see small globules lying in a smooth matrix material. Around this convex surface is the thin ring of an etched face. This is what one sees from the extracellular medium. We are looking down on the surface that lies to the left of the double line shown in Fig. 1.1(a). It, too, shows some small globules, somewhat bigger but more sparse than those on the convex surface, embedded in a smooth matrix. (The pale, rather fuzzy surface at the lower left and right edges of the picture is ice, frozen from the extra-cellular medium. The ice is attached to the extracellular face of the membrane that appears in the picture as emerging from beneath the ice layer.) The small globules in Fig. 1.1(b) lie within the center of the double-layered structure in Fig. 1.1(a) and, clearly, also extend out to the extracellular surface.

What are the materials that compose the cell membrane, this thin but complex structure that separates each cell so effectively from the world around it?

1.2. CHEMICAL COMPOSITION OF BIOLOGICAL MEMBRANES

To determine the chemical composition of cell membranes, we must isolate them. To isolate the plasma membrane (the membrane that bounds the cell as opposed to those membranes that bound the intracellular organelles), we first break the cells, bursting them apart by forcing water or bubbles of air into them or else shattering the membranes in a homogenizer. The membranes are then separated from other cell constituents by centrifugation. To prepare membranes from the cellular organelles (mitochondria, chloroplasts, endoplasmic reticulum, lysosomes and endocytic vesicles: see any textbook on cell biology), the organelles themselves must first be separated from other cell constituents by differential centrifugation and their membranes then isolated as for the plasma membrane. Chemical analysis of the purified membranes shows that they are made up of lipids and proteins, together with smaller amounts of associated carbohydrates. The lipids form the smooth matrix of the image seen in Fig. 1.1(b); the proteins are the globules seen in that micrograph. Different membranes vary greatly in their relative proportions of lipids and proteins, ranging from 20% protein in the case of the myelin layers of membranes that surround nerve cells to 75% protein for the inner membrane of the mitochondrion. In general, those membranes that are the most active metabolically have the greatest proportion of protein. The lipids in any cell membrane are a complex mixture (as we discuss below), and the proteins are very varied. Lipids isolated from cell membranes spontaneously form structured aggregates in water. Certain types of these aggregates, the *liposomes,* appear in cross section in the electron microscope rather similar to the double-layered images seen in Fig. 1.1(a) and, in freeze-fracture, as a featureless, complementary pair of plane surfaces. This is the basis for the view that it is the lipid of the cell membrane that forms the matrix, almost uniform in structure, depicted in Fig. 1.1(b). In contrast, the globular structures seen in Fig. 1.1(b) can be shown by more indirect methods to be proteins. [Indeed, they are visible in Fig. 1.1(b) only because of their attachment to *ferritin,* an iron-containing molecule and itself a protein, which can be bound specifically to the membrane proteins.] Biological membranes are thus made of a matrix of lipid molecules into which are inserted proteins. Other proteins may be attached to this fundamental structure.

Figure 1.1(d) depicts the now well-established arrangement of the lipids and proteins in biological membranes.

1.2.1. Membrane Lipids

Figure 1.2 shows the chemical structure of a few of the lipids that have been isolated from biological membranes. Many of these are built on the backbone of a glycerol molecule, esterified at one end by a phosphate residue and at the other two hydroxyls by fatty acids. Lipids of this

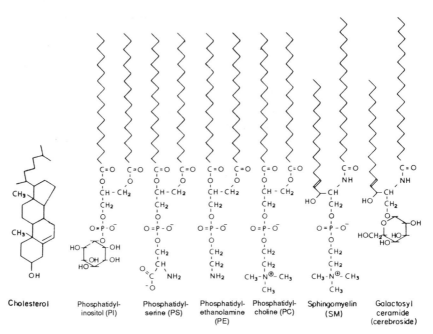

Fig. 1.2.   Some membrane lipids. The zigzag lines represent the hydrocarbon backbone of the lipids. Adapted from J. B. Finean and R. H. Michell, "Membrane Structure." Elsevier/North-Holland Biomedical Press, Amsterdam, 1981.

general class are called *phospholipids*. The charge on the phosphate can be neutralized by its further esterification by a chain bearing an amino group or choline, or the whole molecule can retain a net negative charge. Half the fatty acid chains are saturated; the others carry one, two, or more double bonds. Apart from the fatty acid esters listed in Fig. 1.2, the sterol cholesterol is an important constituent of many animal cell membranes; ergosterol takes its place in plant cell membranes. All of these lipids (and the many other types found as more minor constituents in biological membranes) have the essential physical characteristic of a hydrophilic ("water-loving") portion that interacts strongly with water (the phosphate head-group, in many cases, or the hydroxyl-containing portion of the molecule in some other lipids that are not phosphate esters) and a hydrophobic ("water-hating") portion that inserts into an aqueous environment only with difficulty, if at all (the fatty acid chains and all except the hydroxyl residue of the sterol). Such molecules, which have both a polar and a nonpolar portion, are called *amphiphiles* or *amphipathic*—penetrating both into the water and into the nonaqueous media.

1.2.2. Membrane Proteins

Many membrane proteins are firmly bound to the membranes and can be removed only by treatment with detergents. These are the "intrinsic" proteins. The fact that the intrinsic proteins require detergent treatment to release them from cell membranes suggests that, within the lipid matrix, they are bound to the lipid hydrocarbon chains. The detergent, in fact, replaces the hydrocarbon of the lipid. Other proteins are more loosely attached to the cell membrane. They can be removed by treatment with solutions of low ionic strength, often containing EDTA (ethylenediaminetetraacetic acid), to chelate divalent cations. In red cells, for instance, where half the weight of the dried membrane is protein, some one-third of this protein is lightly attached and can be removed by the above treatment. Such loosely bound proteins are known as "extrinsic" proteins. Most extrinsic proteins seem to be held to the membrane by being bonded to the intrinsic proteins.

1.2.3. Membrane Carbohydrates

Membrane carbohydrates are attached to the proteins, forming the glycoproteins, or to some of the lipid classes [the ceramides (Fig. 1.2)], forming the glycosphingolipids. Protein-bound carbohydrate residues are on the extracellular surface of the cell membrane.

1.3. MEMBRANE PHOSPHOLIPID STRUCTURES
   AND THEIR SELF-ASSEMBLY

We mentioned just above that, in water, phospholipids can self-assemble into "liposomes," structures morphologically resembling the membranes of Fig. 1.1(a) and (b) (without the globular structures, which are the proteins). This is just one of the many forms in which phospholipids can be found in an aqueous medium. Liposomes are single-walled or often multiwalled concentric shells of a bilayer structure. In the simplest single-walled case [as depicted in Fig. 1.3(a) and (b)], a phospholipid bilayer encloses a volume of water and is surrounded by water. The head-groups of the phospholipids are immersed in the water; the hydrocarbon chains, being hydrophobic, point away from the water to the center of the bilayer. The liposomes have proved to be excellent objects for transport studies, behaving as useful "models" of biological membranes.

Bilayers are also formed when organic solvent solutions of phospholipid are painted over a small hole drilled in a plastic partition between

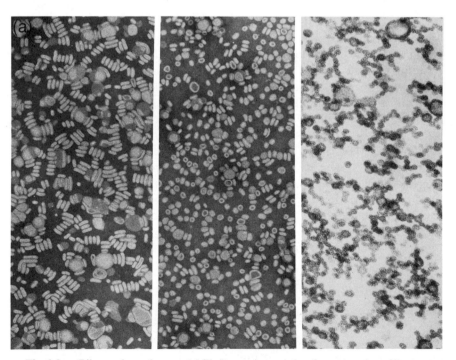

Fig. 1.3. Bilayered membranes. (a) Electron micrographs of a preparation of liposomes, unfixed and unstained. Reprinted with permission from R. L. Hamilton, in "Liposome

(b)

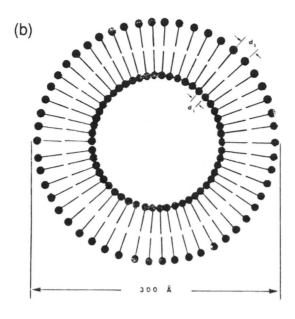

(c)

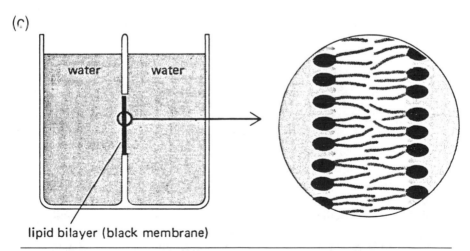

lipid bilayer (black membrane)

Technology,'' Vol. 1 (G. Gregoriadis, ed.). Copyright CRC Press, Inc. Boca Raton, Florida. (b) A diagrammatic sketch of one of these. (Taken from W. D. Stein, ''Movement of Molecules Across Cell Membranes.'' Academic Press, New York, 1967.) (c) *Left:* a sketch of a bilayer formed between two aqueous phases as a black membrane. *Right:* a diagram of the bilayer so formed. Taken, with kind permission, from B. Alberts *et al.,* .''Molecular Biology of the Cell,'' 2nd ed. Garland Publishing, New York, 1989.

two aqueous phases [Fig. 1.3(c)]. The organic solvent dissolves away in the water, leaving a thin film of phospholipid. This gradually thins until it forms a "black" lipid film, so called since the film is so thin that its optical reflectance is very small. The thickness of such a film (about 4–5 nm) is exactly that expected for a bilayer of phospholipids. A bilayer can also be formed by pushing a pierced plastic partition through an aqueous surface on which a surface layer of phospholipid has been spread. At such a water–air interface, a phospholipid spreads so that its hydrophobic groupings point away from the water, up toward the air. As the partition is pushed through the surface layer, the hydrophobic surfaces "zip" together to form the bilayer depicted in Fig. 1.3(c).

The bilayer structure, with the head-groups of the phospholipids immersed in the water and the phospholipid hydrocarbon chains pointing away from it, is energetically the most favorable situation for amphiphilic molecules such as the phospholipids. Water molecules are at a far lower free energy when surrounded by other water molecules with which they can hydrogen-bond, than when abutting onto hydrocarbon chains. Thus water will be excluded from the region of the hydrophobic hydrocarbon chains. These molecules will then line up with one another to form bilayers (or micellar structures also satisfying these thermodynamic requirements in some cases). It is the square, block shape of the phospholipid molecules that encourages the formation of the extended bilayers of Fig. 1.3. (Soaps and lysolecithin, for instance, are wedge-shaped and are more favorably placed energetically in globular structures, although mixtures containing wedges of different shapes can sometimes pack together to form extended bilayers.)

It appears that phospholipids are asymmetrically disposed across the membrane. In the human red blood cell, phosphatidylcholine (PC) and sphingomyelin (SM) are the major phospholipids at the outer face of the membrane whereas phosphatidylethanolamine (PE) and phosphatidylserine (PS) predominate on the inside of the bilayer. In the kidney cell, SM is predominantly on the inside, however. The finding of *asymmetry* is very general and found in a variety of cell types, but there are as yet no general principles pointing to the biological requirement for an asymmetric distribution of any particular phospholipid. This asymmetry is a dynamic one. Physical studies show that phospholipids in red blood cells indeed flip from one face of the bilayer to the other, although seldom. A half-time ($t_{1/2}$) of 8–27 hr has been measured for this flipping rate in human erythrocytes. Even this slow rate should, of course, distribute phospholipids equally between the two faces of the membrane during the 120-day life span of the human red cell in the body. (In contrast to the phospholipids, cholesterol flips across the red cell membrane very fast. An upper bound of 3 sec, at 37°C, for this flipping rate has been reported). It has been established that the asymmetric distribution of the phospholipids in red cells is under metabolic control. We shall see in the next section that proteins, too, are always inserted asymmetrically into the membrane. Perhaps this asymmetry helps to control the asymmetry of the phospholipids.

1.4. MEMBRANE PROTEINS: THEIR STRUCTURE AND ARRANGEMENT

We have seen that many membrane proteins are "intrinsic" proteins, firmly embedded into the phospholipid bilayer [Fig. 1.1(b) and (d)]. Some of these proteins span the membrane, extending all the way across it, from one aqueous face to the other. Others extend only halfway across the bilayer, embedded either in the cytoplasmic face or in the extracellular face of the membrane. These different architectural motifs are associated with different functions of the proteins. The proteins that span the membrane include those that are involved in transmembrane transport of their substrates, while those that extend only halfway across the membrane are often concerned with carrying signals across the cell membrane from the exterior to the interior of the cell.

The first biochemical evidence that proteins span the cell membrane was found by Mark Bretscher. He saw that a major protein of the human red blood cell, a protein now known as *glycophorin* (Fig. 1.4), could be labeled by a nonpenetrating chemical reagent (formylmethionylsulfone methyl phosphate) either when this reagent was present outside the cell or when it was at the cytoplasmic surface of the cell. Hence this protein, glycophorin, extends all the way across, that is, spans, the red cell membrane. Many other proteins have since been shown to span cell membranes. Some of these, like glycophorin, span the membrane only once; others span it many times as the polypeptide chain of the protein winds back and forth between the two faces of the membrane. Let us now consider these membrane-spanning proteins in more detail.

1.4.1. Proteins That Span the Membrane Once Only

Research on the membrane proteins has been accelerated by the use of methods that allow the determination of the *amino acid sequence* of these molecules. (For a description of such methods, see any general textbook on cell biology or biochemistry. For the coming discussion, bear in mind that the end of a polypeptide chain that bears the terminal amino group is

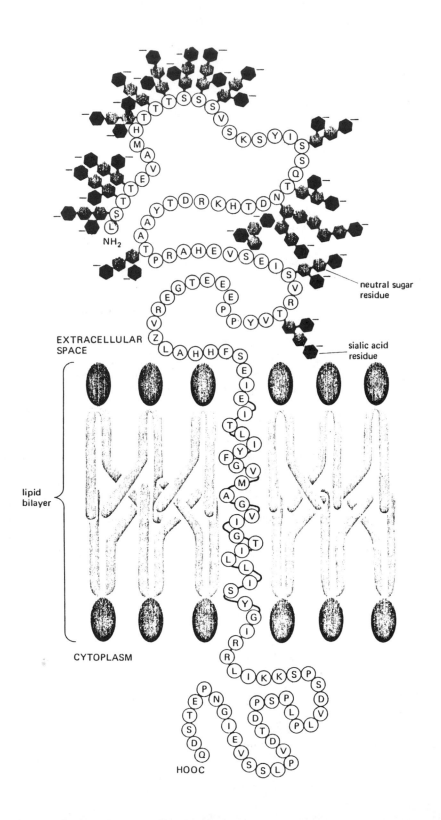

EXTRACELLULAR SPACE

neutral sugar residue

sialic acid residue

lipid bilayer

CYTOPLASM

NH₂

HOOC

termed the *N-terminal end,* while the end bearing the terminal carboxyl is termed the *C-terminal end.* Recall, too, the structure of the alpha (α) helix, one of the fundamental architectural motifs of proteins.) Figure 1.4 depicts the amino acid sequence of the protein glycophorin, which, as we have just seen, spans the membrane once only. When the protein was originally sequenced, it was immediately noticed that a long, 23-membered section of the polypeptide consists of a run of amino acids bearing only hydrophobic side chains. Such a section is long enough that it could insert *as an α helix,* 3 nm long, thus spanning the cell membrane. The amino acid side chains that project from the α helix are able to make hydrophobic contacts with the hydrocarbon chains of membrane phospholipids. In contrast, the amino acids shown in Fig. 1.4 with attached sugar residues bear the carbohydrates of this glycoprotein and, as we mentioned in Section 1.2, these project into the extracellular surface. Thus it seemed very reasonable to suppose that the protein extends across the membrane. Its N-terminal sequence lies in the extracellular medium, a single hydrophobic sequence is buried within the hydrophobic interior of the membrane and the C-terminal portion of the protein is present within the cytoplasm. It is now clear that such a general structure is valid also for a number of other intrinsic membrane proteins.

1.4.2. Proteins That Span the Membrane More Than Once

A second class of intrinsic membrane proteins includes most of the proteins involved in the transfer of material across the membrane (which we shall discuss in great detail in subsequent chapters of this book). Not in every case has it been firmly established that such transport proteins do indeed span the membrane and span it more than once, but it seems fair to generalize such data as are available to include all of this class. We shall consider here three of these multiply spanning proteins in some detail. Others are referred to in later sections of the book.

1.4.2.1. BACTERIORHODOPSIN

Much is known about the protein called *bacteriorhodopsin,* found in the membranes of bacteria, such as *Halobacterium halobium,* which live in

---

Fig. 1.4. Glycophorin inserted into a bilayer. Sketch of how glycophorin may be inserted into a lipid bilayer. The amino acid chain is depicted as a set of letters, being the single-letter codes that symbolize the individual amino acids (listed in Appendix 1). Attached to it at the extracellular surface are sugar and sialic acid residues as depicted. Taken, with kind permission, from B. Alberts *et al.,* "Molecular Biology of the Cell," 2nd ed. Garland Publishing, New York, 1989.

conditions of extremely high natural salinity. When these bacteria find themselves in conditions where oxygen is lacking, they synthesize a "purple" membrane, a part of their cell membrane, which is composed of bacteriorhodopsin molecules arranged in a two-dimensional crystalline array. The protein absorbs light energy and uses this energy to pump protons across the cell membrane. (The process is beginning to be understood, as we shall discuss in Section 6.9.) Bacteriorhodopsin has a relative molecular mass of 26,000 daltons (Da), and has been sequenced. The fact that the purple membrane is crystalline enabled researchers to apply modern methods of crystal structure analysis to determine the structure of bacteriorhodopsin. The analysis by Richard Henderson and Nigel Unwin of the crystals suggested that the protein was built up of seven helical sections, depicted as the cylinders in Fig. 1.5(a), each section spanning the membrane. The amino acid sequence data on bacteriorhodopsin are perfectly consistent with this model and suggest the arrangement of the polypeptide chain as drawn in Fig. 1.5(b). Here, seven segments, each 20–25 residues long, largely composed of amino acid residues with hydrophobic side chains, can be expected to span the cell membrane as α helices.

This concept, that 20–25-member long sequences of particular amino acid residues will form helical rods within a membrane, was generalized by Jack Kyte and Richard Doolittle to provide a most important general scheme for suggesting which regions of known sequence in membrane proteins can be assigned to the membrane-spanning sections.

Each amino acid side chain is given a "hydropathy index"—a number that reflects the frequency with which this side chain is found either buried within a (water-soluble) protein or on its surface. Isoleucine, for example, is most frequently found buried within proteins and its hydropathy index has been set at 4.5. In contrast arginine, most often on the surface, is given an index of −4.5. Stretches of sequences along a protein are analyzed by computer, using a moving-segment approach, and the hydropathy index of the middle point of each segment is plotted against the amino acid sequence number. Short, hydrophobic sequences would be likely to be within the body of the protein, long sequences would be likely to span the membrane (although such assignments are tentative).

1.4.2.2. LACTOSE PERMEASE

Figure 1.6 shows such a hydropathy plot for a second membrane transport protein, the lactose permease that we shall discuss further in Section 5.3.6. The hydrophobicity of each amino acid side chain is plotted on the y-axis (hydrophobic residues being given a plus sign) against the position in the sequence given on the x-axis. Charged residues (positive or

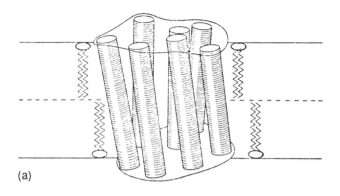

(a)

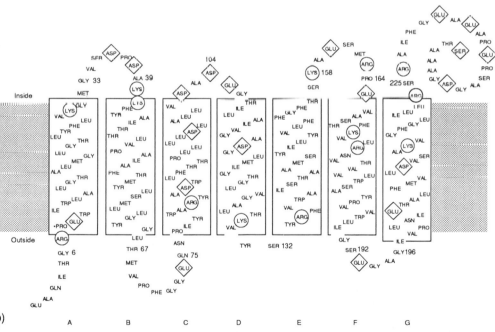

Fig. 1.5.   The structure of bacteriorhodopsin. (a) A schematic view of how the seven α helices of bacteriorhodopsin might be arranged within the bilayer. Taken, with kind permission, from R. Henderson, *Ann. Rev. Biophys. Bioeng.* **6**, 87–109 (1977). Copyright 1977 by Annual Reviews Inc. (b) Possible arrangement of the transmembrane helices of bacteriorhodopsin as deduced from the hydropathy plot (see text). Taken, with kind permission, from D. M. Engelman *et al.*, *Proc. Natl. Acad. Sci. U.S.A.* **77**, 2023–2027 (1980).

LACTOSE CARRIER PROTEIN

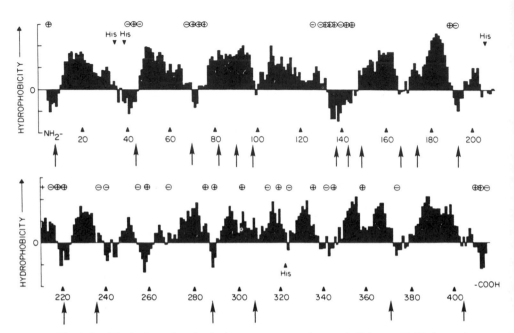

Fig. 1.6. The hydropathy plot for lactose permease (see text). Taken, with kind permission, from D. L. Foster, M. Boublik, and H. R. Kaback, *J. Biol. Chem.* **258,** 31–34 (1983).

negative), which would not normally be expected to be within hydrophobic regions of the membrane, are indicated by circles in the diagram. Twelve main regions which could form helical segments spanning the membrane can be seen. The assignment of any segment to a hydrophobic, helical section is tentative, but this assignment is consistent with measurements of the physical properties of the lactose permease. Such measurements show that 85% of the protein is in an α-helical form, close to the amount that Fig. 1.6 would predict for this molecule. Thus the principle of a multiply spanning polypeptide chain found for bacteriorhodopsin might well be applicable also to the lactose permease.

1.4.2.3. PROTEINS OF THE REACTION CENTER COMPLEX

The reaction center complex of the purple photosynthetic bacterium *Rhodopseudomonas viridis* is the first protein-containing membrane species to be prepared in the form of three-dimensional crystals and to have its structure solved in detail. Figure 1.7 shows a model of its structure as determined by Johann Diesenhofer, Hartmut Michel, and colleagues. The complex consists of three protein species, L, M, and H (for light, medium, and heavy), the protein cytochrome, several quinones, chlorophyll and pheophytins (depicted as the squarish shapes on Fig. 1.7), all molecules concerned in the production of energy from light. The chlorophyll

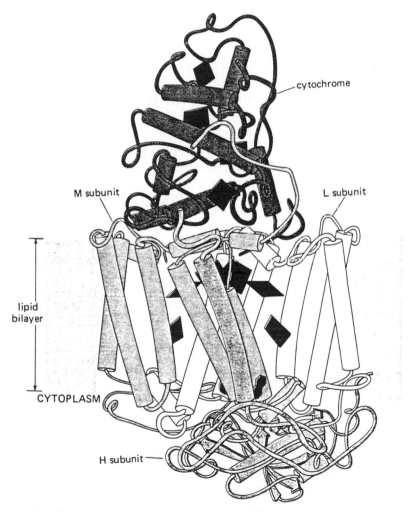

Fig. 1.7.   The reaction center. The structure of the photosynthetic reaction center of the bacterium *Rhodopseudomonas viridis* as determined by X-ray diffraction analysis of crystals of the protein complex. The four subunits of the protein are labeled L, M, H, and cytochrome. The squareish shapes represent the coenzymes. Taken, with kind permission, from B. Alberts *et al.,* "Molecular Biology of the Cell," 2nd ed. Garland Publishing, New York, 1989.

absorbs a photon of light in the first event of a chain in which the energy of the photon drives an electron from the chlorophyll to a pheophytin, from there to a quinone, and on to a second quinone. This second quinone leaves the complex, an event that leads to the buildup of an electric charge across the membrane and to the reduction of the cytochrome. As the structure of the proteins in the complex is now known to a resolution of 0.23 nm, the positions of the amino acids that make up the chain of the proteins have been determined. This has enabled a precise test to be made of the power of the structure–prediction methods of Kyte and Doolittle, described just above. It turned out that the method worked exceedingly well. As can be seen from Fig. 1.7, parts of the chains of the L and M subunits span the membrane and those parts bear those very amino acid residues that the Kyte–Doolittle algorithm predicts should be in membrane-spanning helices. Each subunit, L and M, contains five such helices, arranged in a symmetric fashion around the central axis of the complex. The bulk of the H protein and the cytochromes are arranged as extrinsic proteins, except that one strand of H is membrane-spanning. Embedded in the L–M complex are the chlorophylls and some of the quinones. It is remarkable that such a complex structure could be made to crystallize and the determination of its three-dimensional structure at this high resolution was a wonderful accomplishment.

1.5. SYNTHESIS OF MEMBRANE PROTEINS

We give a very brief treatment of this important subject here. Comprehensive information can be found in cell biology textbooks.

It is now clear that many plasma membrane proteins are synthesized directly into the membranes in which they are found. An intrinsic membrane protein is often characterized by a particular N-terminal amino acid sequence, the so-called signal sequence. This sequence ensures that the ribosome that is in the process of synthesizing it [Fig. 1.8(a)] becomes

---

Fig. 1.8. Synthesis and turnover of membrane proteins. (a) Scheme for synthesis of membrane proteins on endoplasmic reticulum (ER)-attached ribosomes. The mRNA (messenger ribonucleic acid) molecule is depicted as the strip with 3′,5′ ends. It attaches to the ribosome in the cytosol, and the signal polypeptide immediately synthesized then directs the ribosome to the endoplasmic reticulum across which synthesis continues. Taken, with kind permission, from B. Alberts *et al.,* "Molecular Biology of the Cell," 2nd ed. Garland Publishing, New York, 1989. (b) Major routes for membrane traffic, depicted here as in an epithelial cell. The arrows labeled 1 represent the secretory pathways; those labeled 2, endocytosis; while 3 represents synthesis of nuclear and mitochondrial proteins. Taken, with kind permission, from W. H. Evans and J. M. Graham, "Membrane Structure and Function." IRL Press at Oxford University Press, Oxford, England, 1989.

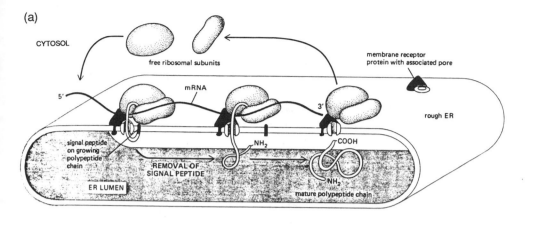

(a)

CYTOSOL

free ribosomal subunits

membrane receptor
protein with associated pore

mRNA

5'

3'

rough ER

signal peptide
on growing
polypeptide
chain

REMOVAL OF
SIGNAL PEPTIDE

NH₂

COOH

NH₂

mature polypeptide chain

ER LUMEN

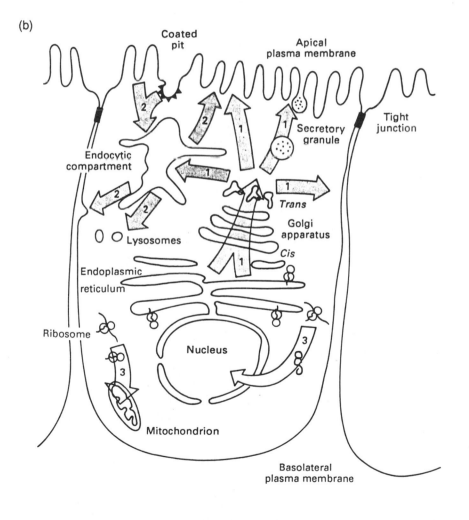

(b)

Coated
pit

Apical
plasma membrane

Tight
junction

Endocytic
compartment

Secretory
granule

Lysosomes

Trans

Golgi
apparatus

Cis

Endoplasmic
reticulum

Ribosome

Nucleus

Mitochondrion

Basolateral
plasma membrane

attached to the membrane of the endoplasmic reticulum. (A ribosome is a complex of membrane proteins and nucleic acids on which occurs the synthesis of the proteins of the cell. The endoplasmic reticulum is a cellular organelle involved in the synthesis of proteins that will be exported from the cell or else inserted into its membranes.) The polypeptide chain that is being synthesized is inserted into the membrane of the endoplasmic reticulum, which afterward, together with the inserted protein, eventually fuses with the plasma membrane of the cell [Fig. 1.8(a) and 1.8(b)]. While the protein is being synthesized on the now membrane-attached ribosome, its polypeptide chain will cross the membrane and the newly synthesized sequences will find themselves outside the cell. A protein destined to be an intrinsic membrane protein bears, in addition to its signal sequence, a sequence or sequences consisting of a 20–25-member-long chain of hydrophobic residues, which will become and remain embedded in the membrane lipids. A protein that spans the membrane once only always appears to have its N-terminal sequence outside the membrane, a direct consequence of this mode of synthesis where the membrane-spanning sequence appears after a longer or shorter sequence of membrane-leaving residues. A protein that spans the membrane a number of times will possess a set of hydrophobic sequences. Each sequence remains within the membrane, and the polypeptide chain "sews itself" into the membrane as this thread of the polypeptide chain doubles back on itself during synthesis. (A protein destined for export will, apart from the signal sequence, bear the amino acid sequence typical of a soluble protein, with no long, hydrophobic sequences.) The proteins that compose the mitochondrion and chloroplast are inserted into the membranes of these structures by a more complex process. Most of them are coded for by nuclear genes and some have to cross the outer organellar membranes to reach their final insertion point. A few are coded for by organellar genes and some of these, again, reach their sites by a complicated route.

The lipid and the intrinsic proteins of the membrane undergo a more or less rapid *turnover* [Fig. 1.8(b)]. In many cells, parts of the plasma membrane invaginate as a whole, forming sacs that are coated with a protein called *clathrin*. These "coated pits" break off from the membrane as vesicles. They fuse with existing cytoplasmic organelles (the endocytic vesicles or endolysosomes) and eventually with the lysosomes of the cell in which the membrane constituents are digested. Certain specific proteins present in the coated pits, indeed the majority, escape this incorporation into lysosomes and return to the membrane from which they came [Fig. 1.8(b)].

Extrinsic membrane proteins (like cytoplasmic proteins, but unlike intrinsic membrane proteins) are synthesized on ribosomes that are present in the cytoplasm as soluble and not membrane-associated organelles. The extrinsic proteins make their way to the membrane, presumably by diffusion and associate there with their intrinsic, specific counterparts.

1.6. PHASE TRANSITIONS IN BIOLOGICAL MEMBRANES

Membrane phospholipids are packed in a regular array, head to head, shoulder to shoulder, and tail to tail. Such arrays are crystalline but can be ordered (true crystals) or disordered (liquid crystals). Figure 1.9 is a diagram showing the order-to-disorder transition in an array of phospholipid molecules arranged in a bilayer. The ordered bilayer "melts" as the temperature is raised through a critical temperature range (ΔT in Fig. 1.9) over which the phase transition occurs. A protein molecule present in such a bilayer may well display quite different kinetic properties when it is surrounded by phospholipids in an ordered, in contrast to a fluid, state. It is important to establish the state of the phospholipids of the biological membranes with which we will be concerned. Are they fluid or ordered at physiological temperatures? It turns out that the fluidity of a membrane depends on the nature of the phospholipids of which it is made. For a fluid membrane, the requirements are short chains and much unsaturation. The requirements for a solid bilayer are long chains and little unsaturation,

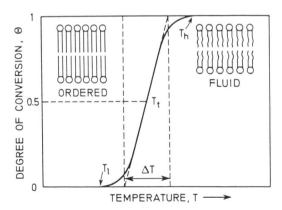

Fig. 1.9.   The phase transition in lipid bilayers. The diagram represents the melting of an ordered (solid) bilayer on the left to a fluid one on the right. The membrane "melts" over the temperature interval ΔT. Taken, with kind permission, from P. Overath and L. Thilo, *in* "Biochemistry of Cell Walls and Membranes II" (J. C. Metcalfe ed.), pp. 1–44. University Park Press, Baltimore, Maryland, 1978.

where the hydrocarbon chains will pack closely together and interact strongly.

To see what sort of effects can arise as the phospholipids in a membrane undergo a phase transition, consider some fascinating experiments that Peter Overath and colleagues performed. They grew bacteria under various nutritional conditions in which quite different phospholipids were forced to become inserted into the bacterial membranes. They were thus able to vary the transition temperatures of the phospholipids in the membranes of these cells over the range from 15° to 35°C! Bacteria with these different membrane characteristics were then grown over a wide temperature range. It was shown that the bacteria grew successfully at all temperatures at which the phospholipids were in the fluid state (no ordered phospholipid is necessary to ensure growth!), but could grow even when as much as half of the phospholipid was ordered. Above this degree of order the bacteria grew with increasing difficulty as the degree of ordering increased. Many bacteria and even higher organisms can adjust the membrane hydrocarbon chain composition of their membranes according to the growth temperature. The characteristic length of the hydrocarbon chains found in common organisms is 16 or 18 carbon atoms, while the degree of unsaturation commonly found is one double bond per two chains. These values seem to be a compromise between the requirement for a fluid membrane and the requirement for a stable bilayer.

1.7. QUANTITATION OF MEMBRANE DYNAMICS

We have seen that most of the phospholipids in cell membranes are in a fluid state at physiologic temperatures. One can, however, improve on this qualitative statement and express quantitatively the dynamics within the fluid membrane. We find that there is a "flexibility gradient" down the length of the hydrocarbon chain. The head-groups and glycerol backbones (see Figs. 1.2 and 1.3) in even a fluid phospholipid bilayer are highly ordered. If one performs measurements that can resolve movements on a time scale of 10^7 to 10^8 events per second, one finds that the head-groups are scarcely free to move. The flexibility of the chain, however, increases about 200-fold as one moves toward the center of the bilayer. The tails of the hydrocarbon chains move very freely. Membrane proteins appear to order the phospholipids that are found in their immediate environment, forming a ring of "immobilized" lipid around the protein molecule. Up to 55 moles of phospholipid are immobilized by each mole of the protein cytochrome oxidase when this is added to a bilayer. Addition of intrinsic proteins to phospholipid bilayers has the effect of increasing the viscosity of the fluid bilayer.

The proteins and phospholipids in bilayers can be shown to be capable of rotating. The protein rhodopsin, the light-absorbing protein present in the retina of the eye, rotates approximately 60° every 10 microseconds (μsec). This value is consistent with the fact that the membrane has a viscosity about 200 times the viscosity of water, the membrane being well modeled by the viscosity of a light oil.

Proteins and phospholipids can also move sideways (laterally) from place to place in the plane of the membrane. Typically, a *phospholipid* in a liposome will diffuse about 50–100 nm laterally in a millisecond. In a cell membrane, where the proteins, as we have just seen, stabilize the lipids, the rate of movement is only some two-thirds of this. The rates of lateral movement of *proteins* vary over a very large range. The fastest, rhodopsin in retina, can cover 20 nm in a millisecond, a value consistent with the speed with which it rotates and consistent also with the relative sizes of the phospholipid and rhodopsin molecules. An important membrane protein, which we will see in Section 4.8.1, is responsible for anion transport in red blood cells, diffuses some thousand times more slowly than does rhodopsin. Yet measurements of its ability to rotate showed that it could rotate freely in the membrane. Such results strongly suggest that there are severe constraints to the free lateral movement of many proteins in the plane of the cell membrane, but not to their rotation (nor to the lateral movement of the lipids). We now know that below the cytoplasmic surface of the membrane is a highly organized structure, the cytoskeleton, made up of an extended array of specific proteins (see Fig. 1.10, which depicts the cytoskeletal–membrane protein interactions beneath the membrane of the red blood cell). It is the interaction between the membrane proteins and the proteins of the cytoskeleton of the cell that anchors the membrane proteins at particular places in the plane of the membrane.

We shall see in later chapters of this book that many cells, especially those concerned with absorption of metabolites, are vectorially *polarized* in that they have quite different membrane protein compositions at two opposite faces of the cell. Clearly the membrane proteins are held in place in the membrane, within a certain "fenced-off" area. Here, too, it is most likely that these restraints to lateral diffusion involve transient interactions of the intrinsic membrane proteins with other (extrinsic) proteins present on the cytoplasmic surface of the membrane.

1.8. THE CELL MEMBRANE AS A BARRIER AND AS A PASSAGE

The preceding discussion of membrane structure is intended to provide a framework for our subsequent extended discussion of the movement of

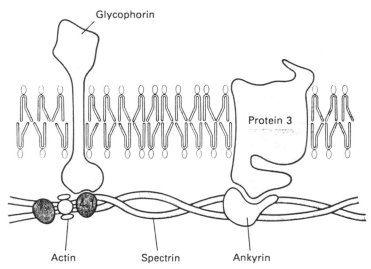

Fig. 1.10.   Interactions between membrane proteins and the cytoskeleton. The diagram shows two intrinsic membrane proteins, glycophorin and "Protein 3" (the anion transporter), interacting with elements of the cytoskeleton that lies beneath the cell membrane. Seen are the long chains of spectrin and the proteins ankyrin, "4.1," and actin that interact with spectrin and with membrane proteins, or with each other, to form a supercomplex. Taken, with kind permission, from W. H. Evans and J. M. Graham, "Membrane Structure and Function." IRL Press at Oxford University Press, Oxford, England, 1989.

ions and molecules across biological membranes. What does structure tell us about the barrier function of membranes and about how passages may be found across these barriers? To summarize, it seems that biological membranes are essentially bilayers of phospholipids into which are inserted, in many cases, molecules of cholesterol and proteins. Such compound bilayers form a continuous sheet around the cell and around the individual cellular organelles. For a molecule to enter or leave the cell or organelle, it must cross this sheet.

We should perhaps deal separately with the barrier and passage properties of the lipid and protein components of the cell. If a molecule is to cross the membrane, it must first enter the lipid region, then cross it, and leave at the other face. We have seen that the lipids are almost always fluid. The head-groups pack tightly together, but the hydrocarbon tail regions are very flexible. To enter the lipid region is to enter an hydrophobic environment. Any bonds that the would-be permeant makes with the aqueous medium bathing the membrane will have to be broken before the permeant can leave the water and enter the hydrophobic portion of the membrane. A charged group such as an ion will have great energetic problems in order to enter the hydrophobic environment. Only very rarely will it enter as a fully charged, naked ion. The major cellular metabolites are almost always charged or bear several hydrophilic groups that form hydrogen bonds with water. Thus, they also will seldom cross the lipid barrier of the cell. This is clearly the major function of the barrier, to keep the products of cellular metabolism within the cell. If a molecule does succeed in getting into the hydrophobic region of the membrane, it must then cross it. It will experience in the region of the head-groups a tightly packed quasi-crystalline matrix presenting a stiff barrier to diffusion, but the center of the bilayer with its highly flexible hydrocarbon tails will be far less of a barrier to intramembrane movement.

What of the protein portion of the membrane? First, we have seen that the protein apparently immobilizes around itself a ring-like domain of lipid. Functionally, this ring probably seals the intrinsic proteins within the lipid, maintaining the highly hydrophobic character of the interior of the cell membrane right up to the surface of the intrinsic protein. What are the barrier properties of the proteins themselves? We can say little about this as yet. The globular, water-soluble proteins that have been analyzed structurally have densely packed interiors, mostly hydrophobic in nature. Thus they would have the barrier properties of phospholipids, probably to an even higher degree of effectiveness. But we shall see in later chapters that membrane proteins may well be constructed so as to form a hydrophilic channel along their central axis. These proteins might form passages enabling selected ions or molecules to cross the membrane. As we shall see, other proteins most probably act as selective transporters of their specific substrates by opening up specific gates or doors for their substrate molecules.

Because the phospholipids are fluid in nature, they are able to allow passage of hydrophobic substances. A fluid membrane might well be mandatory, also, for the protein conformational changes that we shall show in later chapters to be the basis of many transport processes. How the phospholipids act as barrier and passage will be the subject of much of Chapter 2 of this book. How some transmembrane proteins can act as channels will be the content of Chapter 3, while the more complicated highly specific transport proteins will be dealt with in Chapters 4–6. Meanwhile, in the first half of Chapter 2 we shall consider some of the physical principles that govern movement across cell membranes.

SUGGESTED READINGS

General

Alberts, B. *et al.* (1989). "Molecular Biology of the Cell," 2nd ed., Chapter 6. Garland Publishing, New York.

Bretscher, M. S. (1985). The molecules of the cell membrane. *Sci. Am.* **253,** 100–109.

Finean, J. B. R., Coleman, R., and Michell, R. H. (1984). "Membranes and Their Cellular Functions," 3rd ed. Blackwell, Oxford.

Membrane Structure

Findlay, J. B. C., and Evans, W. H., eds. (1987). "Biological Membranes: A Practical Approach" IRL Press, Oxford.

Klausner, R. D., Kempf, C., and van Renswoude, J., eds. (1987). "Membrane Structure and Function" Curr. Top. Membr. Transp., Vol. 29. Academic Press, San Diego, California.

Kleinfeld, A. (1987). Current views of membrane structure. *Curr. Top. Membr. Transp.* **29,** 1–27.

Membrane Lipids

Deamer, D. W., and Uster, P. (1985). Relation of liposomes to cell membranes. *In* "Structure and Properties of Cell Membranes. III. Methodology and Properties of Membranes" (G. Benga, ed.), pp. 103–122. CRC Press, Boca Raton, Florida.

Edidin, M. (1987). Rotational and lateral diffusion of membrane proteins and lipids: Phenomena and function. *Curr. Top. Membr. Transp.* **29,** 91–127.

Vassileo, P. M., and Tien, H. T. (1985). Planar lipid bilayers in relation to biomembranes. *In* "Structure and Properties of Cell Membranes. III. Methodology and Properties of Membranes" (G. Benga, ed.), pp. 63–101. CRC Press, Boca Raton, Florida.

Membrane Proteins

Ragan, C. I., and Cherry, R. J., eds. (1986). "Techniques for the Analysis of Membrane Proteins." Chapman & Hall, London.

Unwin, P. N. T., and Henderson, R. (1984). The structure of proteins in biological membranes. *Sci. Am.* **250,** 56–67.

Membrane Dynamics

Edidin, M. (1987). Rotational and lateral diffusion of membrane proteins and lipids: Phenomena and function. *Curr. Top. Membr. Transp.* **29,** 91–127.

Quinn, P. J., Joo, F., and Vigh, L. (1989). The role of unsaturated lipids in membrane structure and stability. *Prog. Biophys. Molec. Biol.* **53,** 71–103.

Glycophorin

Bretscher, M. (1973). Membrane structure: Some general principles. *Science* **195,** 743–753.

Bacteriorhodopsin

Henderson, R. *et al.* (1990). Structure of bacteriorhodopsin at 0.3 nm resolution. *J. Mol. Biol.* (in press).

Stoeckenius, W., and Bogomolni, R. A. (1982). Bacteriorhodopsin and related pigments of Halobacteria. *Annu. Rev. Biochem.* **51,** 587–616.

Lactose Permease

Kaback, H. R. (1983). The *Lac* carrier protein in *Escherichia coli*. *J. Membr. Biol.* **76,** 95–112.

Hydropathy Plots

Guy, H. R. (1985). Amino acid side-chain partition energies and distribution of residues in soluble proteins. *Biophys. J.* **47,** 61–70.

Kyte, J., and Doolittle, R. F. (1982). A simple method for displaying the hydropathic character of a protein. *J. Mol. Biol.* **157,** 105–132.

Photosynthetic Reaction Centre

Deisenhofer, J. *et al.* (1985). The structure of the protein subunits in the photosynthetic reaction centre of *Rhodopseudomonas viridis* at 3 Å resolution. *Nature (London)* **318,** 618–624.

Synthesis of Membrane Proteins

Alberts, B. *et al.* (1989). "Molecular Biology of the Cell," 2nd ed., Chapter 8. Garland Publishing, New York.

Endocytosis, Membrane Turnover

Alberts, B. *et al.* (1989). "Molecular Biology of the Cell," 2nd ed., Chapter 8. Garland Publishing, New York.

Hare, J. F. (1990). Mechanisms of membrane protein turnover. *Biochim. Biophys. Acta* **1031,** 1–69.

Cytoskeleton

Bennet, V. (1985). The membrane skeleton of human erythrocytes and its implications for more complex cells. *Annu. Rev. Biochem.* **54,** 273–304.

Carraway, K. L., and Carraway, C. A. C. (1989). Membrane-cytoskeleton interactions in animal cells. *Biochim. Biophys. Acta* **988,** 147–171.

CHAPTER 2

Simple Diffusion of Nonelectrolytes and Ions

In this chapter we will describe how molecules and ions move across cell membranes by *simple diffusion,* that is, unaided by any specific component of the membrane. We first consider, in a very general way, the forces that act on a molecule or ion so as to cause its movement across cell membranes. We then go on to consider what properties of the cell membrane determine the rate at which such a molecule or ion crosses a particular membrane.

It appears that many substances of great importance to the cell are associated with special systems that speed up their movement above the rate at which they would cross an otherwise unmodified cell membrane. (Such systems often act on these particular ions or molecules so as to bring about their concentration within the cell, or their exclusion from it.) These are the *mediated transport* systems, in which transport is mediated by some specific component of the membrane, and are the concern of Chapters 3–7. This chapter, in contrast, is concerned with *nonmediated transport.*

2.1. DIFFUSION AS A RANDOM WALK

We know from experience that a vaporous substance liberated in a corner of a room, or a solute initially confined to one part of a container [Fig. 2.1(a)], will eventually come to be distributed over the whole volume accessible to it [Fig. 2.1(b)]. This diffusion is brought about by the ceaseless, random movements of the diffusing molecule, and of the particles

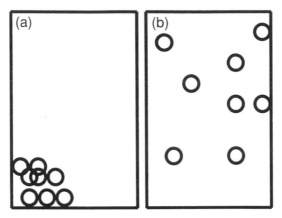

Fig. 2.1.   Diffusion as a random walk. (a) The initial state. All the molecules of the substance S are confined to one corner of the box. (b) The final, equilibrium state after the random walk has taken place (and is still continuing). The molecules are now more or less evenly distributed over the entire volume of the container.

composing the medium in which it finds itself. If two compartments, in only one of which a substance, S, is originally present [Fig. 2.2(a)], are separated by a barrier that S can cross, S will again by these ceaseless, random movements find itself equally distributed over the whole volume accessible to it, now both containers [Fig. 2.2(b)]. The molecules are ceaselessly colliding with the barrier. Therefore, at any moment, the number of molecules of S crossing the barrier from side I to side II will be proportional to the number impinging on the barrier (and hence to the concentration of S in the solution at side I), multiplied by the area, A, of

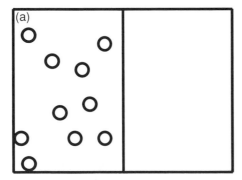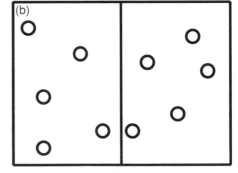

Fig. 2.2.   A random walk across a permeable barrier. (a) The initial state. All the molecules are confined in the left-hand compartment. (b) The final state after the molecules have taken a random walk that took them across the barrier between the two compartments. The molecules are now evenly distributed over the entire volume of the two containers.

the barrier. The number crossing from side II to side I will, similarly, be proportional to the concentration of S at side II. The *net* movement from side I to side II will be determined by the *difference in concentration* of S in the two bathing solutions. (*Note:* We use "concentration" here where, strictly, "activity" should be used. The concept of activity takes into account the interactions that occur between the molecules of a solute and reduce its "effective" concentration. At low concentration these interactions become of little significance and the activity of a solute approaches its concentration. For most biological situations, the two are sufficiently close together for us to ignore their difference.)

The other factor that determines the rate of transbarrier movement is the *resistance* that the barrier offers to the movement of S across it. Factors influencing the resistance are the nature of the barrier, whether it is loosely textured or tightly packed, and the strength of the physicochemical forces of interaction between the diffusing molecule and the molecules that compose the barrier. A tightly packed barrier that avidly binds the diffusing molecule will impede the random movements of the diffusant and reduce the rate at which it can cross the barrier. A high barrier resistance leads to a low transbarrier diffusion rate. Clearly also, the thickness of the barrier is a determining factor, in that the thicker the barrier, the more it impedes the progress of the diffusant from the one solution to the other.

Putting these intuitive concepts into a mathematical form, we can see that the rate of net transbarrier movement is inversely proportional to the resistance of the barrier ($\mathbf{R}$) and to its thickness (d), but directly proportional to the concentration difference between sides I and II and to the area of the barrier (A). Writing $J_{\text{I}\rightarrow\text{II}}$ for the rate of net movement of S from side I to side II, and S_{I} and S_{II} for the concentrations of S at sides I and II of the barrier, respectively, we obtain

$$J_{\text{I}\rightarrow\text{II}} = A(S_{\text{I}} - S_{\text{II}})/\mathbf{R}d \qquad (2.1)$$

[Equation (2.1) is strictly true only if the barrier is homogeneous. If, instead, it varies in its properties along the diffusion path, one has to integrate Eq. (2.1) over the thickness of the barrier.]

It is usual in studies of diffusion to define the reciprocal of the resistance of the barrier as D, its *diffusion coefficient* with respect to the substance S. Resistance and diffusion always have this reciprocal relationship with each other. Sometimes, as in electrical studies, it is more convenient or more usual to use resistance; in other circumstances its reciprocal, diffusivity, may be more convenient to use. (In Section 3.1, we shall meet with conductances, also reciprocals of resistances, hence directly proportional to diffusion coefficients).

TABLE 2.1

Some Diffusion Coefficients[a]

Substance	Molecular/ionic weight	Temperature (°C)	Diffusion coefficient (D) ($\times 10^{-5}$)
H_2	2	21	5.2
O_2	32	18	2
Na^+	23	25	1.33
K^+	39	25	1.96
Cl^-	35.5	25	2.03
I^-	127	25	2.04
SO_4^{2-}	96	25	1.06
Water	18	20	2.1
Urea	60	20	1.2
Glucose	180	20	0.6
Lactose	342	20	0.43
Raffinose	504	20	0.36
Insulin	12,000	20	0.15
Myoglobin	17,500	20	0.11
Hemoglobin	68,000	20	0.063
Urease	490,000	20	0.034
Tobacco mosaic virus	4×10^7	20	0.0053

[a] Values are for diffusion in water in dilute solutions. Data for D in cm^2 sec^{-1}.
The data were taken from tables compiled by R. Höber (1947) "Physical Chemistry of Cells and Tissues" Churchill, London, and F. Daniels and R. A. Alberty (1961) "Physical Chemistry" Wiley, New York.

Box 2.1. Einstein's Relationship between D and $t_{1/2}$

In the early 1900s Albert Einstein showed that there is a simple relation between the diffusion coefficient, D, of a substance and the time, t, that it takes to diffuse an *average* distance, d, in a fluid medium. The relation for diffusion in one dimension is $d^2 = 2Dt$. Now for glucose diffusing in water at room temperature, D is 6×10^{-6} cm^2 sec^{-1} (see Table 2.1 for a listing of diffusion coefficients). This means that glucose takes about 0.08 sec to diffuse across an intestinal cell [of length 10 micrometers (μm)]. The protein hemoglobin (of diffusion coefficient 6×10^{-7} cm^2 sec^{-1}) and its substrate oxygen (of diffusion coefficient of 2×10^{-5} cm^2 sec^{-1}) take just under 1 sec and some 25 msec, respectively, to diffuse the length of the red blood cell (also some 10 μm). These are short times, and it does not seem that intracellular diffusion rates will be limiting for the life of these cells. Note,

(Continued)

however, that distance comes into Einstein's relation as a square. This means that for the very large protozoan cell *Paramecium,* which has a length of 100 μm, all of these values will have to be increased 100-fold. For such large cells, the rate of intracellular diffusion might well become a limiting factor, reducing the effectiveness of the cell. One is entitled to think, therefore, that the constraints of diffusion might set an upper limit to the size of biological cells. For a nerve cell, which can be meters in length, substitution in Einstein's relation shows that a tremendously long time (hundreds of years!) would be required if diffusion were the only mechanism by which substances traveled inside these cells. Specialized intracellular transport systems have developed to cope with this problem.

Equation (2.1) can be rewritten in terms of diffusion coefficients as

$$J_{I \to II} = DA(S_I - S_{II})/d \qquad (2.2)$$

where D is the diffusion coefficient of S across the barrier that separates I and II. Equation (2.2) is a form of Fick's first law of diffusion and is the basis of any quantitative understanding of diffusion and membrane transport. [The coefficient D has the dimensions $cm^2 sec^{-1}$. Check this out in Eq. (2.2) by substituting the appropriate units for J (moles per second), A (cm^2), S (mol/cm^3), and d (cm).]

Clearly, from Eq. (2.2), net flow of S across the barrier will occur until the right-hand side of Eq. (2.2) is zero, that is, until $S_I = S_{II}$, when the concentration difference across the barrier disappears. Thus, diffusion continues until the system has reached the condition of maximum disorder, when S is uniformly distributed between the two compartments. (At this condition of maximum disorder, the *entropy* of the system has reached a maximum. Refer to any textbook on physical chemistry regarding this topic and the other concepts that we will handle in the next few paragraphs.) Recall that, when the system has reached equilibrium, the *chemical potential* of S is the same at the two sides of the barrier. The chemical potential of a substance is a measure of its capacity to perform work.

We shall need to use some mathematical symbolism in the next paragraph and the few subsequent pages. It is, however, well worthwhile patiently working through this material carefully since the results derived are of such great importance to a full understanding of transport processes.

Let us denote U_j as the chemical potential of the solute S_j, U_j° as its standard-state chemical potential (i.e., its chemical potential at a concentration of one molal [i.e., 1 mol of S_j per 1000 grams of water] at 0°C, and at zero electrical potential, $\overline{V}_j$ as its partial molal volume (i.e., the increment in volume per mole of solute added when an infinitesimal amount of S_j is added to the solution), P as the pressure exerted on the solution (in excess of atmospheric pressure), ψ as the electrical potential at which S_j finds itself, and finally Z_j as the valence of the ion j. We then can write the expression

$$U_{j,i} = U_j^\circ + RT \ln S_{j,i} + \overline{V}_j P + Z_j F \psi_i \qquad (2.3)$$

where the subscript i stands for either I or for II, indicating which side of the barrier is being referred to. This expression is true for a substance at sea level, when no work is done against the gravitational force. (The gravitational component of the chemical potential, which is not treated further here, is important in considering the water relations of plants that often have to move water many meters upward against the force of gravity.) The Faraday constant F is the electrical charge carried by 1 mol of univalent ion (again, refer to any textbook on physical chemistry regarding the Faraday constant). Why the electrical potential is important will be seen in Section 2.2.

Equation (2.3) is most important, and we shall use it a good deal in the following chapters. It displays clearly the sources of the four kinds of work that a chemical substance can perform, if it is provided with a means of doing work. The first source, its standard-state chemical potential, can be mobilized as work if the substance in question is transformed into another of a different standard-state potential. This is the *chemical work* that can be performed by or on the substance during a chemical change. The second source arises from the concentration of the substance. As the substance in question moves from a higher to a lower concentration, it can perform work. Movement from a lower to a higher concentration requires work. This term is the concentration, or *osmotic work*. Work can also be done by expanding the volume of the system against an external pressure (*pressure work*). This third component of the total work is often of small importance in biological systems, but determines the phenomenon of osmotic pressure. (See Box 2.4, where this point is taken up further). The fourth source is the *electrical work*. As a charged substance is moved in an electric field, from one electrical potential to another, it can perform work or have work done on it, depending on its charge and the direction of the field.

These forms of work are interchangeable—one into the other. All are, of course, fundamental in the life of the cell. Chemical work is ceaselessly being performed during cellular metabolism, osmotic work is being done on or by solutes that cross cellular and intracellular membranes, and pressure work is done when water is moved across these membranes, while the electrical potential that exists across all cell membranes is the background for the transmembrane movement of charged solutes. The interaction between these sources of energy is the subject of much of this present book.

2.2. THE ELECTRICAL FORCE ACTING ON AN ION

If our substance, S, is an ion, there is, in addition to the concentration gradient, an extra force acting on it to bring about transport and that is the electrical potential difference across the barrier (see Fig. 2.3). This elec-

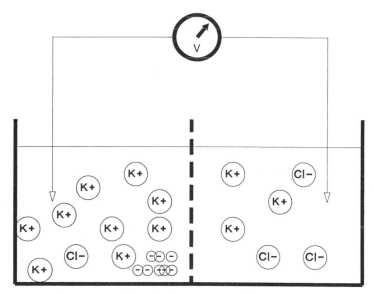

Fig. 2.3. Origins of the equilibrium (Nernst) potential. Two aqueous compartments are separated by a membrane barrier but connected by a wire to a voltmeter, V, which measures the potential difference across the membrane. The left-hand compartment contains an impermeable anion, symbolized by the small, connected circles. Both compartments contain potassium ions (the large circles enclosing a K^+) and chloride ions (the large circles enclosing a Cl^-). At equilibrium the net charge is almost zero in each compartment so that there is an excess of cations in the left-hand one, to balance the charge on the impermeable anion. The tendency of the permeable cations to flow down their concentration gradient is balanced by the membrane potential, which draws them back into the left-hand compartment. In the case of the permeable anion, the same basic equilibrium of forces ensures that they are at a higher concentration in the compartment that *does not* contain the impermeable anion.

trical potential is a real push-or-pull force against which an ion will have to diffuse, or in the direction of which its diffusion rate will be accelerated. The word "ion" comes from the Greek for "wanderer" and reminds us that these substances move under the action of an applied electrical force. The ion feels the fields of force of the electrical field that is being applied across the barrier, and its diffusion is enhanced or retarded accordingly.

To find the ion distribution at equilibrium, we use Eq. (2.3) for the chemical potential, remembering that, at equilibrium, the system can no longer do any work against either the chemical or the electrical gradient. (We can combine these two together in the single concept of the *electrochemical* gradient.) The chemical potential is, therefore, the same in the two phases. Thus we can equate the right-hand sides of two versions of Eq. (2.3), one where the side $i = I$ and the other where $i = II$. Realizing that the standard-state chemical potential is the same for the ion at the two sides of the barrier, we find in solving the resultant equation that

$$RT \ln S_I + ZF \, \psi_I = RT \ln S_{II} + ZF \, \psi_{II} \qquad (2.4)$$

which, on writing $\Delta\psi$ for the difference in electrical potential across the barrier, rearranges to

$$\Delta\psi = RT/ZF \ln(S_{II}/S_I) \qquad (2.5)$$

We shall refer to Eq. (2.5) many times in subsequent discussions. It shows how, at equilibrium, the electrical potential difference balances the osmotic potential across the membrane. The difference $\Delta\psi$ is measured as the potential at side I minus that at side II. (It is important to get these conventions of sign correct. They cause a good deal of confusion if they are not used consistently. To check the convention, note that for a cation, positively charged, $\Delta\psi$ is positive if the ion concentration at side II is greater than that at side I. Thus side I is more positive than side II, pushing the cation away from side I toward side II, just as we find at this equilibrium.)

In terms of logarithms to the base 10, Eq. (2.5) becomes

$$\Delta\psi = 2.303RT/ZF \log(S_{II}/S_I) \qquad (2.6)$$

The term $2.303 \, RT/ZF$ works out at 59 mV, at 25°C (=298 K) and for an ion of unit valence ($Z = 1$). The constant $R = 1.987$ cal K^{-1} mol^{-1} (8.314 J K^{-1} mol^{-1}) and $F = 23,060$ cal V^{-1} mol^{-1} (96,500 J V^{-1} mol^{-1}). (We have written K for degrees kelvin, J for joules and V for volts. Values of the fundamental constants used in this book are listed in Appendix 2.) Thus a univalent ion is at equilibrium when a tenfold concentration ratio ($S_{II}/S_I = 10$) balances a 59-mV transbarrier electrical potential difference, a useful number to remember. Many animal cell membranes are found with a ten- to 50-fold higher concentration of potassium within the cell than outside it. This will balance more than a 59 mV transmembrane potential, inside negative, not far from the values often found. (A 100-fold concentration difference—10^2—will balance out a $2 \times 59 = 118$ mV potential, etc.)

TABLE 2.2

Ion Concentration Ratios, Calculated Equilibrium Potentials, and Measured Transmembrane Potentials[a]

Cell type	Ion	Concentration ratio (out/in)	Potentials		Ref.[b]
			Equilibrium (calculated)	Membrane (measured)	
Muscle (mammalian skeletal)	Na⁺	12	+67	−90	(i)
	K⁺	0.026	−98		
	Ca²⁺	15,000	+123		
	Cl⁻	30	−90		
Muscle (embryo heart)	Na⁺	9	+57	−70	(ii)
	K⁺	0.040	−83		
	Ca²⁺	12,000	+120		
	Cl⁻	5.1	−42		
Axon (squid)	Na⁺	8.5	+55	−60	(iii)
	K⁺	0.029	−91		
	Ca²⁺	10,000	+118		
	Cl⁻	14	−67		
Red blood cell (human)	Na⁺	18	+74	−10 to −14	(iv)
	K⁺	0.05	−77		
	Ca²⁺	52,000	+139		
	Cl⁻	1.6	−12		
Algal cell (*Nitella*)	Na⁺	0.07	−67	−138	(v)
	K⁺	0.0008	−179		
	Cl⁻	0.02	+99		

[a] Potentials in millivolts, calculated from Eq. (2.6) or measured.

[b] The data were compiled from various sources (see Suggested Readings) as follows: (i) Hille (1984, p. 14); (ii) M. Lieberman et al. 1984, in Blaustein and Lieberman, eds., p. 184; (iii) R. A. Sjödir (1984), in Blaustein and Lieberman, eds. p. 121; (iv) J. C. Ellory and V. I. Lew, eds., "Membrane Transport in Red Cells," Academic Press, London, 1977 (data from various articles therein); and (v) Nobel (1974, p. 105).

Table 2.2 lists ion concentrations and transmembrane potentials for a variety of cell types.

Equation (2.5) shows that, at equilibrium, the *concentrations* of an ion, S, at the two faces of the barrier will not be the same if there is a net transbarrier electrical potential (see Fig. 2.3). Equation (2.5), known as the *Nernst equation,* defines the Nernst or *equilibrium potential* for an ion, the potential difference for which it will be in equilibrium across the barrier if it is at the concentrations S_I and S_{II} at sides I and II of the barrier. The force on the ion brought about by the transbarrier electrical potential difference is here just balanced against the diffusive "force" due to the concentration difference of the ion across the barrier. (The origin of the transmembrane potential found in all living cells will be discussed in Section 3.7.) Clearly, ions of opposite signs will have to be at oppositely poised concentration ratios if they are both to be at equilibrium. To check this, first put $Z = 1$ and then $Z = -1$ in Eq. (2.5) and compare the resulting ion distributions for the same $\Delta\psi$. Thus with a 59-mV transbarrier potential, side I positive to side II, a cation will have to be at a 10-fold higher concentration at side II than at side I to be in equilibrium, while an anion will have to be at a 10-fold lower concentration at side II than at side I for it also to be in equilibrium. In many animal cells, the chloride and potassium ions have such oppositely poised concentration ratios [with potassium high and chloride low, intracellularly (see Table 2.2)], both ions often being not far from equilibrium with the electrical potential across the membrane. If the concentration gradient for S is not that given by Eq. (2.5), the system has the capacity to perform work, which it will do if the means of doing so are provided.

It should not be thought that, at equilibrium, there is no flux of the ions across the membrane. Although there will be no net flux, there will indeed be a constant traffic of ions in both directions across the membrane, the flux in the two directions being equal and opposite. Box 2.2 deals with the relation between the transmembrane fluxes when the system is not necessarily at equilibrium.

Equation (2.4) is, of course, also valid for noncharged molecules, since it is derived from Eq. (2.3), which is true for both uncharged and charged solutes. To check this, substitute zero for Z (as the valence of the nonelectrolyte) into Eq. (2.3) and then rederive Eq. (2.4). We see, of course, that for any value of $\Delta\psi$, the transbarrier potential, $S_I = S_{II}$ at equilibrium. Thus sugars and amino acids bearing no net charge will be at equilibrium across the cell membrane when their transmembrane concentration ratios are unity, at any transmembrane potential. This will not be true for amino acids bearing a net charge, when Eq. (2.5), with the appropriate value of Z, will determine the equilibrium distribution.

Box 2.2. Ussing's Flux Ratio Test

The Danish physiologist, Hans Ussing, introduced a convenient relationship that describes the fluxes of ions across a membrane as a function of the transmembrane potential, $\Delta\psi$, and of the ion concentrations at the two sides of the membrane, S_I and S_{II}. It is clear from our discussion of Fig. 2.2 that the flux from a particular face should be proportional to the concentration of the ion at that face. In addition, our intuition about electrical fields enables us to realize that the flux of a positive ion from the side at low potential will be reduced while that from the side at high potential will be increased above the flux in the absence of a potential difference. But if we do not know how the electrical field is distributed *through* the membrane, we cannot directly apportion some particular part of it to the increase or decrease in a particular flux. What we can do, however, is to relate the transmembrane potential difference, $\Delta\psi$, to the *ratio* of the ion fluxes in the two directions. Writing the fluxes in the directions I to II and II to I in terms of the J notation used in Eqs. (2.1) and (2.2), we can put for the flux ratio of an ion of valence Z:

$$J_{I \to II}/J_{II \to I} = (S_I/S_{II})\exp(ZF\,\Delta\psi/RT) \qquad (B2.2.1)$$

[At equilibrium, the fluxes across the membrane are equal and their ratio is thus unity. We regain Eq. (2.5).]

Equation (B2.2.1) will be obeyed whenever the fluxes are carried by the ions without their being coupled to any source of energy other than that provided by the transmembrane field. If there is any such coupling, or if the ion fluxes are carried on some system for which the flux is not directly proportional to concentration (such as will be the case for fluxes on the mediated systems that we shall discuss in Chapters 3–7), Eq. (B2.2.1) may not hold. Behavior contrary to that predicted by Eq. (B2.2.1) suggests that the ion fluxes are moving by some system that is not a simple one, but rather one that is coupled to an energy input or else mediated in some other fashion.

2.3. PERMEABILITY COEFFICIENTS AND PARTITION COEFFICIENTS

Our formulation of the law of transbarrier diffusion [Eq. (2.2)] contains two terms that are seldom possible to measure in biological membranes:

d, the thickness of the barrier; and D, the diffusion coefficient *across* the barrier. The thickness of a biological membrane can, it is true, be measured from electron micrographs [see Fig. 1.1(a) and refer to any cell biology textbook], but this is the overall thickness and we do not know the thickness of that part of the membrane that is the major barrier to diffusion. As we have seen, biological membranes are not uniform in their structure as one traverses them in the direction from, say, side I to side II. Which of the various regions (the head-groups, the tightly packed first half of the hydrocarbon chains, or the more loosely packed hydrocarbon tails) most strongly holds back a diffusing permeant? Until we know the answer to this question, it is safest to take an overall view and define the *permeability, P,* of the membrane to a particular substance as the ratio of two directly measurable quantities, the measured transmembrane flow $J_{I \to II}$ and the measured concentration difference ($S_I - S_{II}$). Referring to Eq. (2.2), we see that $J_{I \to II}$ divided by $S_I - S_{II}$ is merely DA/d. We can now replace the difficult to define parameters D and d by their ratio D/d and call this the permeability, P, the ratio of net flow to the force that drives this flow, the concentration difference.

We obtain, by substituting in Eq. (2.2), a fundamental equation that defines the permeability coefficient:

$$J_{I \to II} = P_s A (S_I - S_{II}) \tag{2.7}$$

where P_s is the permeability coefficient of S across the membrane. Thousands of measurements have been made of the permeability coefficients for many different substances, mostly non-electrolytes, crossing membranes of various types of cells. Table 2.3 records a small fraction of these data. In Section 2.4, we discuss how such permeability coefficients are measured. Meanwhile, let us go a little more deeply into the factors that determine what the permeability coefficient will be for a certain substance crossing a particular cell membrane.

Equation (2.7) is exact and we can always use it to determine the permeability coefficient of a membrane for a particular substance. But what if we are obstinate and, for all our hesitations, want to estimate the diffusion coefficient of this substance *within* the material of which the membrane is composed? Can we do this? The answer is that we can, indeed, make an educated guess as to the intramembrane diffusion coefficient, and the values we obtain tell us a good deal about the barrier properties of cell membranes.

Our problem is that Eq. (2.7) is defined in terms of the concentration difference of the substance S in the media bathing the faces of our barrier (see Fig. 2.2). To measure diffusion *within* the barrier, we need to know

TABLE 2.3

Permeability Coefficients for Various Compounds Crossing Membranes of a Range of Cell Types[a]

Permeant	Permability coefficients			
	Chara ceratophylla (algal plant cell)	Human red blood cell	Urinary bladder (toad)	Artificial lipid bilayers
Water	6.6×10^{-4}	1.2×10^{-3}	1.1×10^{-4}	2.2×10^{-3}
Formamide	2.2×10^{-5}	1.1×10^{-6}		1.0×10^{-4}
Butyramide	5.0×10^{-5}	1.1×10^{-6}	5.0×10^{-6}	
Urea	1.1×10^{-6}	7.7×10^{-7}	1.0×10^{-6}	4.0×10^{-6}
Thiourea	2.0×10^{-6}	1.1×10^{-6}		
Ethanol	1.6×10^{-4}	2.1×10^{-3}	1.0×10^{-4}	
Ethanediol	1.1×10^{-5}	2.9×10^{-5}	8.2×10^{-7}	8.8×10^{-5}
Glycerol	2.0×10^{-7}	1.6×10^{-7}		5.4×10^{-6}
Erythritol		6.7×10^{-9}		

[a] Permeabilities in cm sec^{-1}, and determined at 25°C.
The data for *Chara* are from R. Collander and H. Bärlund (1933) *Acta Bot. Fenn.* **11**, 1–114; for human red blood cells from various authors as tabulated by W. R. Lieb and W. D. Stein, in W. D. Stein (1986) "Transport and Diffusion across Cell Membranes" Academic Press, Orlando; for toad urinary bladder from N. Bindslev and E. M. Wright (1976), *J. Membr. Biol.* **81**, 159–170; and those for artificial lipid bilayers from E. Orbach and A. Finkelstein (1980) *J. Gen. Physiol.* **66**, 251–265.

the concentration difference of S *within* the barrier, and this will, in general, be quite different from ($S_I - S_{II}$). The membrane constitutes a chemical phase quite different from that of the media that bathe it, so that its solution properties are quite different. Hydrophobic substances, such as the higher alcohols and hydrocarbons, will have a far greater tendency to dissolve in the lipid of the membrane than in the aqueous media bathing it. Conversely, hydrophilic substances, such as sugars and amino acids, will dissolve far more readily in the aqueous media, and will be more or less excluded from the membrane interior. The measure of the relative tendency of a substance to dissolve in a solvent as compared with its tendency to dissolve in water, is given by its solvent–water *partition coefficient,* defined as the ratio of the solubility in the chosen solvent to the solubility in water, at the temperature at which the measurement is made. The partition coefficient is generally given the symbol K. It is most conveniently measured by the apparatus depicted in Fig. 2.4.

Here, a separatory funnel holds a sample of the substance S, whose partition coefficient is to be measured, some water, and some of the chosen solvent (which must be immiscible with water). The funnel is

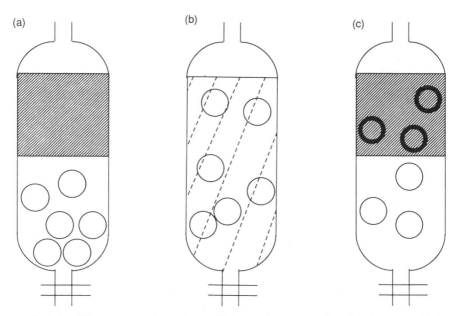

Fig. 2.4. Measurement of partition coefficients. A separatory funnel is shown to which, in (a), molecules dissolved in the lower, aqueous phase have been added with an organic phase added above the aqueous phase. In (b) the two phases are well mixed in the funnel. In (c), they have been left to separate out again, when the added molecule is seen to be distributed between the two phases. These can readily be sampled and the concentrations of the dissolved molecule in each phase measured. The partition coefficient is the ratio of these two concentrations.

shaken for many hours until S is at equilibrium between the two phases, and then left to stand until the two phases separate out again. The concentrations of S in the upper and lower phases are determined. The ratio of the concentration in the solvent to that in water gives the partition coefficient, K. Clearly, both solvent and substance must be specified for K to be properly defined. Table 2.4 records some partition coefficients obtained using a number of different organic solvents.

With this concept of partition coefficient in mind, consider Fig. 2.5, which depicts a concentration profile that might be found for a membrane situated between two bathing solutions. (This is really too simplified a picture for a biomembrane, which we saw in Chapter 1 to be composed of a number of regions differing with respect to hydrophobicity and, hence, solubility for a diffusing molecule. The treatment that follows lumps together all these different regions of the membrane in terms of an average

TABLE 2.4

Partition Coefficients for the Distribution of Various Solutes Between Water and Different Solvents[a]

Solute	Partition coefficients		
	Hexadecane	Olive oil	Octanol
Water	4.2×10^{-5}	1.3×10^{-3}	4.1×10^{-2}
Formamide	2.2×10^{-5}	1.1×10^{-6}	
Butyramide	5.0×10^{-5}	1.1×10^{-6}	5.0×10^{-6}
Urea	2.8×10^{-7}	1.5×10^{-4}	2.2×10^{-3}
Thiourea		1.2×10^{-3}	7.2×10^{-2}
Ethanol	5.7×10^{-3}	3.6×10^{-2}	4.8×10^{-1}
Ethanediol	1.7×10^{-5}	4.9×10^{-4}	1.2×10^{-2}
Glycerol	2.0×10^{-6}	7.0×10^{-5}	2.8×10^{-3}
Erythritol		3.0×10^{-5}	1.2×10^{-3}

[a] At 25°C or room temperature.
The data were originally tabulated by W. R. Lieb and W. D. Stein, in Stein (1986).

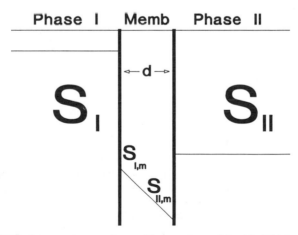

Fig. 2.5.   Diffusion across a membrane. The membrane (Memb) of thickness d separates two phases I and II, containing the permeable solute S which is at concentration S_I in phase I and S_{II} in phase II. *Within* the membrane, there is a concentration gradient of S that is at concentration $S_{I,m}$ just within the membrane bathing phase I and concentration $S_{II,m}$ just within the membrane at phase II. The thin lines represent the levels of concentration of the solute S.

partition coefficient. Indeed, it turns out that it is the region with the *lowest* partition coefficient that provides the rate-determining barrier to diffusion.) The concentration gradient that drives the diffusion of the solute *within* the membrane is the concentration difference $S_{I,m} - S_{II,m}$ where this expression refers to the concentrations *in* the membrane adjacent to the bathing solution.

At either face of the membrane, the permeant S will be in equilibrium across the interface between the membrane interior and the adjacent bathing solution. The partition coefficient K for this substance across the membrane–water system defines the concentration ratio between the bathing solution and the membrane adjacent to it. If S_I and S_{II} are the concentrations of S in the bathing solutions and $S_{I,m}$ and $S_{II,m}$ are the concentrations within the membrane, our partition coefficient gives us that

$$K = S_{I,m}/S_I = S_{II,m}/S_{II} \tag{2.8}$$

We know the concentrations in the bathing media, and we can use Eq. (2.8) to calculate the concentrations in the membrane, adjacent to the bathing solutions. These are KS_I and KS_{II}, adjacent to sides I and II, respectively. Rewriting Eq. (2.2) in terms of the diffusion events taking place within the membrane, and defining D_{mem} as the diffusion coefficient of S *within* the membrane, we obtain

$$J_{I \to II} = D_{mem}A(KS_I - KS_{II})/d \tag{2.9}$$

or

$$J_{I \to II} = D_{mem}AK(S_I - S_{II})/d \tag{2.10}$$

where the concentrations refer to the aqueous solutions bathing the membrane.

Compare Eqs. (2.7) and (2.10). They both refer to the same flow of S and the same concentrations of S, those in the aqueous solutions, and to the same membrane area, A. Hence it follows that the remaining terms in the two equations can be equated. We have, therefore, that

$$P_s = D_{mem}K/d \tag{2.11}$$

This is a most important equation. It shows us that the readily measured membrane permeability coefficient, P_s, is determined by three parameters: the diffusion coefficient of the permeant within the membrane, the membrane–aqueous partition coefficient of this permeant, and the thickness of the membrane barrier. We shall use Eq. (2.11) in Section 2.6 to estimate values of the membrane–water partition coefficients for various substances crossing different cell membranes, and values of their intramembrane diffusion coefficients. This will give us much insight into the nature of the cell membrane as a permeability barrier.

2.4.   MEASUREMENT OF PERMEABILITY COEFFICIENTS

Meanwhile, we must consider how one *measures* permeabilities. Equation (2.7) tells us what we have to do to make such measurements: we have to measure the *rate* at which S crosses the membrane, the *concentrations* of S in the two solutions bathing the membrane, and the area, A. Measurement of a rate entails measurement of the *amount* of S that has entered or left the cell during a known period of *time*.

We generally measure amounts chemically, or radioactively using labeled permeants, or by fluorescent measurements with fluorescently labeled permeants. We can also measure amounts indirectly by measuring the volume change of the cell as water enters or leaves it, following the movement of the permeant. (In Section 2.7 we will consider in detail such linked flows of water and permeant, when we consider the phenomenon of *osmosis*.)

Measurement of time implies that we can identify the instant that the cells and permeant are mixed together, and the instant that the permeant ceases to cross the membrane. In the simplest such setup, often used by cell biologists [Fig. 2.6(a)], the cells, attached to a plastic tissue culture dish, are washed briefly and then a solution of the permeant is added at zero time. At some later time (seconds to hours!), the solution is sucked off the dish, the dish washed rapidly in ice-cold solution free of permeant and the amount that has entered the cells is determined.

More sophisticated methods, for fast-moving permeants, use a rapid-mixing device, whereby cells in suspension (often preloaded with permeant) are mixed with the permeant solution by forcing them through a mixing chamber. In one version of this method [Fig. 2.6(b)], the mixture flows along a tube from which the suspension medium is filtered across Millipore filters inserted at intervals into the tube. The fluid emerging from the filter has been in contact with the permeant-loaded cells for a defined time before the separation. Thus the amount of permeant that has left the cells during that time can be estimated. Alternatively, in this and other systems (for instance, the liposomes discussed as models for cell membranes in Section 1.3), the cells can be mixed in a second mixing chamber with an ice-cold solution free of permeant to slow down further transport. The cells are then filtered off at leisure and their content of permeant determined. This stopping solution can contain inhibitors of the transport system when the measurement is for one of the specific transport systems to be discussed in later chapters.

In another technique, frequently used for cells that are in, or can be brought into, suspension (blood cells or tumor cells), the cells, after the desired period of incubation, can be centrifuged through an organic phase, intermediate in density between the cells and the aqueous solution. The cells sediment, but the lighter aqueous solution remains as a supernatant. Thus, at the moment the centrifuge is switched on, the cells are removed from contact with the aqueous medium, and the movement of S

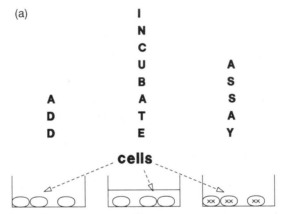

Fig. 2.6. Some methods of measuring membrane permeabilities. (a) Using cells attached to a tissue culture dish: on the left the solution in which the cells have been cultured is removed, and in the center a solution containing, generally, radioactively labeled substrate is added to the dish and left in contact with the cells for a measured time. Thereafter, on the right, this solution is removed, these cells washed rapidly several times while still attached to the dish, and the labeled substrate (xx) that has entered the cells is extracted by breaking up the cells. The radioactivity so extracted is measured. A knowledge of the specific activity of the ratioactive label (the counts associated with each mole of substance) enables us to calculate the amount of substrate that entered each cell in unit time. If the surface area of the cell is known, the permeability of the membrane to the solute in question can be calculated using Eq. (2.7). (b) Using a rapid-mixing device: two syringes are connected via a mixing chamber to a long tube, pierced at various point along its length. One (the upper) syringe is filled with a suspension of red blood cells, preequilibriated with a solution of the (radioactively labeled) permeant. The other syringe contains the washout fluid. After mixing, the now-diluted red cell suspension flows along the tube and the extracellular fluid is expelled across Millipore filters embedded into the successive outflow ports. The expelled fluids from the different outflow ports are collected over a period of time and their radioactivity measured. Successive distances along the tube correspond to successive times after dilution of the red cell suspension. Thus, from the rise in radioactivity as one samples the various collecting tubes, one can calculate the permeability of the solute. (c) Using centrifugation across a water-immiscible phase: cells are incubated with a solution

(b)

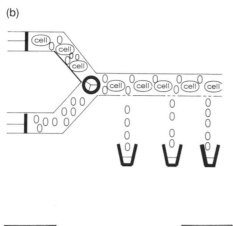

(c)

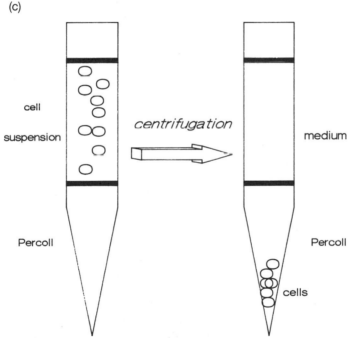

of the permeant for a desired period of time. Permeation is then stopped by layering the cells rapidly on top of a water-immiscible phase (e.g., Percoll, silicone, or dibutyl phthalate) and immediately centrifuging. If the lower phase is of the correctly chosen density, the red cells will sediment through it, leaving behind the aqueous phase with its content of permeant. The cells, freed from the extracellular phase, pack down at the bottom of the centrifuge tube. The tube can be cut with a sharp blade so that the cells can be selectively removed and their content of the permeant measured. From the rise (or loss) in this content with time and a knowledge of the total surface area of the cells, one can calculate the permeability coefficient.

across the cell membrane virtually ceases [Fig. 2.6(c)]. This method can fail if S is too soluble in the organic phase.

Increasingly, measurements of permeability are being made by the techniques of nuclear magnetic resonance (NMR) and electron spin resonance (ESR). For example, the nuclei of the hydrogen atoms of water have a spin which is measured in the NMR spectrometer. This signal is quenched in the presence of manganous ions. If red blood cells are rapidly transferred to a manganous chloride-containing solution that does not cross the cell membrane, the NMR signal from the water outside the cells is quenched and the remaining signal is then dominated by the exchange of water from inside the cell to the outside. Water diffusion across red cell membranes has been measured in this way over a wide temperature range and for a wide distribution of animal species and of inhibitors (see Tables 2.3 and 2.6, where some of these data are collected).

Molecules can be synthesized containing unpaired electrons whose spin can be identified by ESR techniques. The range of these so-called spin probes or labels now include large hydrophobic anions and cations containing the spin probe TEMPO. The electron spin is again quenched by manganous ions so that movement of the probes across the membrane can be studied when cells or liposomes loaded with the probes are transferred to solutions containing manganous chloride. The large size of these probes makes it possible to investigate directly the dependence of transmembrane diffusion on the molecular size of the diffusing molecule (see Section 2.6).

Measurements of the permeability coefficients of very rapidly permeating substances are complicated by the presence of the *unstirred layers* that exist on the surface of all membranes. These have the effect of adding the resistance of a layer of water on each face of the membrane to the resistance offered by the membrane itself. For a substance that moves *extremely* fast across the lipid membrane, the unstirred layers may represent the major barrier to the overall movement from outside to inside the cell or vice versa. Methods are available that allow the complications of the unstirred layer to be overcome or at least identified. Box 2.3 gives a brief introduction to these methods. The thickness of the unstirred layer that remains when even the most vigorous mechanical stirring is applied depends on the physical structure of the cell or tissue concerned. Values that have been reported range from 1 μm for human red blood cells to as much as a millimeter for the deeply folded intestine. As a rough guide, permeability coefficients may be seriously underestimated owing to the presence of unstirred layers if they are higher than 10^{-2} cm sec^{-1} in the case of red blood cells, one-tenth this value for most animal cells, another one-tenth lower for gallbladder and algal cells, and one-tenth lower again (that is, 10^{-5} cm sec^{-1}), for intestinal cells.

Box 2.3. Correcting for Unstirred Layer Effects

The overall resistance to the flow of a substance across a membrane, R_{tot}, on each side of which an unstirred layer is present, is given by the sum of the resistance due to the membrane itself, R_{mem}, and the resistance of the two unstirred layers, which we combine as R_{us}. Thus

$$R_{tot} = R_{us} + R_{mem} \qquad \text{(B2.3.1)}$$

We saw in Section 2.1 that permeability coefficients can be seen as the reciprocals of resistances. Thus we can rewrite Eq. (B2.3.1) as

$$1/P_{tot} = 1/P_{us} + 1/P_{mem} \qquad \text{(B2.3.2)}$$

which we can rearrange to give

$$1/P_{mem} = 1/P_{tot} - 1/P_{us} \qquad \text{(B2.3.3)}$$

Thus, if we know the terms on the right-hand side of Eq. (B2.3.3), we can calculate the desired permeability coefficient as the left-hand term. Now, P_{tot} is simply the permeability as measured without taking into account the presence of the unstirred layer. All the methods that we have discussed up to now give us this value of P. The term P_{us} is the permeability of the unstirred layer alone, which we must next find out how to measure.

The permeability of the unstirred layer is found by eliminating, in some way or other, the contribution of the membrane itself to the overall permeability. What then remains is the effect of the unstirred layer only. We can eliminate the effect of the membrane itself by choosing a permeant for which we suspect that *all* the resistance to the overall flow is due to the unstirred layer. Consider a substance which has a very high lipid solubility, such as butanol. From Eq. (2.11), the rate at which this substance will cross the lipid portion of the membrane will be very high. It is reasonable to expect that the membrane itself will not be the major barrier to the overall movement, but rather the unstirred layer. For such a highly lipid-soluble substance, the measured overall permeability will be a good approximation to P_{us}, the permeability of the unstirred layer.

For the human red blood cell, the permeability coefficient of butanol was found to be 6×10^{-2} cm sec^{-1}. We can use this value and Eq. (2.11) to calculate the *thickness* of the unstirred layer, if we assume that for butanol all the resistance to overall movement is contributed by the unstirred layer. For then, butanol is diffusing through the *water* of

(Continued)

the unstirred layer with a diffusion coefficient given by its value for diffusion in bulk water (1×10^{-5} cm^2 sec^{-1}). Substituting this value for D in Eq. (2.11), and the measured permeability coefficient for P, we arrive at an estimate of the thickness of the unstirred layer as 1.7×10^{-4} cm, or 1.7 μm. (K is, clearly, unity for water versus water.)

Another procedure that has been successfully used to make allowance for the unstirred layer is to create conditions so that the membrane itself presents no barrier to the movement of the permeant. Consider the toad urinary bladder. For this system, like others that we shall consider in Section 7.2.2, the antidiuretic hormone (ADH) markedly stimulates water flow across the epithelium. We do not know, from first principles, whether we have to take account of the unstirred layer in such measurements of water permeability. What was done to check this point was to measure water permeability as a function of increasing amounts of ADH. At sufficiently high concentrations of the hormone, no further increase in water permeability could be achieved by the addition of hormone. Thus, at these high levels of ADH, the membrane itself was apparently not the rate-limiting barrier for water movement across the bladder. We can assume that it was the unstirred layer that was now rate-limiting. Thus, a measure of this maximum rate of water permeability was a measure of the effect of the unstirred layer, P_{us} in Eq. (B2.3.3). Using this equation, the permeability of the membrane itself could now be found from measurements of overall flow performed in the absence of hormone, or even at low levels of hormone where the unstirred layer was not yet rate-limiting for the overall transport of water.

With P_{us} and P_{tot} known we can thus calculate P_{mem}, the permeability of the membrane barrier itself.

We have discussed just a few of the many methods that have been used to measure the permeability of a wide variety of cells and artificial membranes to a great range of substances under a variety of conditions such as changes in temperature. Examples of these data are recorded in Table 2.3. In the next section, we go on to analyze some of the data obtained.

2.5. ANALYSIS OF PERMEABILITY DATA

The Finnish botanist, Runar Collander, working in the 1930s, made some of the most valuable measurements of permeability by a wise choice

of material. He took cells of the marine alga *Chara ceratophylla,* which
has very large cells (up to 2 cm in length and 1.2 mm in cross section)
which can be manipulated by hand. Collander took individual cells and,
after washing them free of seawater in which they had been growing,
suspended them in seawater solutions containing a wide range of test
substances, mostly nonelectrolytes, whose permeabilities he wished to
determine. He left the cells in these solutions for various periods of time
(seconds to hours), until significant amounts of the permeants had entered
the cells. Then he washed the cells again and, by accurate chemical methods, determined the amount of the substance that had entered the cell.
The cells were of very regular size and thus lent themselves to an accurate
measurement of the area of surface available for diffusion. As a function
of the time of soaking in the test solution, he obtained data such as that
depicted in Fig. 2.7.

The different permeants enter the algal cells at different rates. Clearly,
ethylene glycol (EG in Fig. 2.7) enters the cell more rapidly than does
lactamide (LA), and this again more rapidly than glycerol (GLY). One can
define a "half-time" for the entry of a test substance into a cell as the time
taken for the internal concentration to reach one-half that of its equilibrium value. For ethylene glycol, the half-time is about 45 min, for lactamide it is 7–8 hr, and for glycerol, 2 days. Mathematical analysis shows
(Appendix 3) that this half-time ($t_{1/2}$) is related to V, the volume of the cell
into which permeation is occurring, A its surface area, and P_s, the permeability of the penetrating substance, by Eq. (2.12):

$$t_{1/2} = 0.693\,V/(P_s A) \qquad (2.12)$$

Thus measurements of the half-time, with V and A known, gave the
permeability of the permeant. In a classic presentation (Fig. 2.8), Collander plotted, on the y-axis, the permeability of 36 substances (including the
three plotted in Fig. 2.7) against the partition coefficients (see Section 2.3
and Table 2.4), on the x-axis, as measured for the solvent olive oil. Since
the data varied over a very wide range, Collander used a logarithmic scale
on each axis. (Actually, Collander multiplied the value of each permeability coefficient by the square root of the molecular mass of the solute and
plotted this product term in Fig. 2.8. He did this in order to make some
correction for the *size* of the permeating molecule. We discuss this question of size again in a moment.)

There is obviously a very good log/log correlation between the permeability of a substance and its partition coefficient. Why should this be, and
what does this tell us about the membrane's barrier to diffusion? To
understand this relation, return to Eq. (2.11), which related the permeability coefficient, P, and the partition coefficient, K. We see that the two

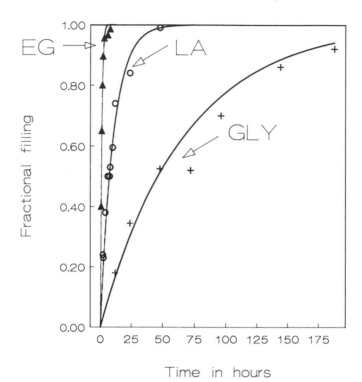

PERMEABILITY OF CHARA
(Collander and Barlund, 1933)

Fig. 2.7. Permeability of three solutes across the plasma membrane of *Chara cera-tophylla*. These are modern replots of some of the classical data of Runar Collander and Hugo Barlund, *Acta Bot. Fenn.* **11,** 1–114 (1933). The data were measured, at room tempera-ture, by direct chemical analysis of the entering solute (EG, Ethylene glycol; LA, lactamide; and GLY, glycerol). The lines are theoretical plots for half-times of 44 min, 7.3 hr and 46 hr, respectively, for the three solutes.

coefficients are directly proportional to each other, related by the factor D/d. [Indeed, $\log P = \log K + \log(D_{mem}/d)$.] Note that the relevant parti-tion coefficient in Eq. (2.11) is the membrane–water partition coefficient (compare Fig. 2.5). If permeability is directly proportional to the olive oil–water partition coefficient (Fig. 2.8), it follows that the diffusion barrier of a membrane has solvent properties similar to those of olive oil. Thus, Fig. 2.8 shows that olive oil seems a good model for the partitioning behavior of the cell membrane of *Chara ceratophylla*!

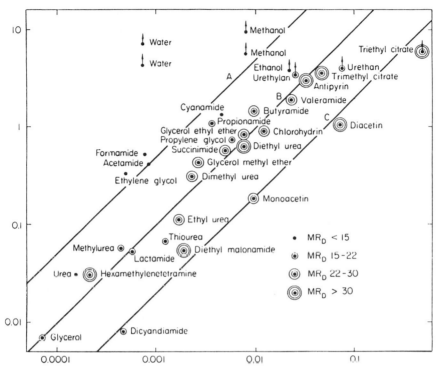

Fig. 2.8. Lipid solubility determines permeability coefficients. The permeability of cell membranes of the plant alga *Chara ceratophylla* to many different organic nonelectrolytes (permeability in cm hr⁻¹, multiplied by the square root of the molecular mass of each substance) is plotted on the ordinate on a logarithmic scale against its olive oil–water partition coefficient, also on a logarithmic scale. The term MR_D is the molar refraction of the permeant, a parameter proportional to the molecular volume. Taken, with kind permission, from R. Collander, *Physiol. Plantarum* **2,** 300–311 (1949).

2.6. THE MEMBRANE AS A HYDROPHOBIC SIEVE

But we can get still more detailed information on the diffusion barrier of a membrane if we think a little more deeply about plots such as Fig. 2.8. In Fig. 2.8 the largest molecules, e.g., diethylmalonamide, are depicted as triple circles, those of intermediate size, e.g., valeramide and propionamide, as double and single circles, while the smallest molecules, e.g., formamide, are shown as dots. We note that, in almost every case, points relating to large molecules fall below the average straight line in Fig. 2.8, points relating to small molecules fall above that line. In addition to the marked effect of the partitioning behavior of the membrane, there seems also to be a sieving effect in that the movement of the larger molecules is hindered. This would be the case if the molecules that constitute the diffusion barrier of the membrane were fairly tightly packed.

Consider also Fig. 2.9, which contains plots, again on a logarithmic scale, of the permeabilities of the human red blood cell membrane for various substances plotted against partition coefficients for each permeant, using hexadecane as the model solvent (see Table 2.4). These are *basal* permeabilities, where all relevant specific transport systems (and red cells have many such specific systems; see Chapter 4) have been blocked by inhibitors. This is, once again, a relatively good straight line. Clearly, permeabilities and partition coefficients go hand in hand, and the partitioning behavior of the red blood cell membrane, like that of *Chara,* is modeled well by an organic solvent. But, again, it appears that large molecules fall below the average line, small molecules above it. Can we obtain a quantitative measure of the sieving effect of the membrane?

Go back, now, to Eq. (2.11). If we divide a permeability value by the appropriate partition coefficient, we obtain the quantity D_{mem}/d, directly proportional to the diffusion coefficient *within* the membrane for the permeant in question. Now it seems sensible to assume that the solvent hexadecane (a long-chain hydrocarbon, very similar in structure and properties to the hydrocarbon chains of the membrane's phospholipids) might be appropriate as a good model for the solvent properties of the cell membrane. We can, therefore, use hexadecane–water partition coefficients (Table 2.4) and Eq. (2.11) to estimate relative values of the intramembrane diffusion coefficients for our series of permeants. [Return to Eq. (2.11) to see how this can be done.] In Fig. 2.9(b) (lower line), we plot these derived diffusion coefficients against the molecular volume of the permeant concerned, on a semilogarithmic scale. We note that the diffusion coefficients fall off very steeply with molecular size (although this effect may level off above a molecular mass of ~100 daltons). The upper line in Fig. 2.9(b) shows a similar plot for diffusion coefficients of molecules diffusing in water. Again, diffusion coefficients decrease with molecular size, but those measured for water depend far less steeply on size than do those we have calculated for diffusion within the cell membrane. The molecules that form the cell membrane and between which intramembrane diffusion has to take place appear to be far more organized structures than is liquid water. Indeed, diffusion through lattices composed of polymers also has the steep dependence on size that we see for diffusion within the membrane.

What gives diffusion within cell membranes this steep size dependence? Return to the discussion of membrane structure in Chapter 1 and recall that the lipid component of the cell membrane is composed largely of phospholipids whose hydrocarbon chains lie more or less at right angles to the plane of the membrane. Between these hydrocarbon chains lies the route along which molecules must cross the cell membrane. The organized structure of the hydrocarbon chains (shoulder to shoulder, tail parallel with tail) is somewhat analogous to the structure of a soft polymer and seems to provide the structural basis for the sieving properties of the cell membrane.

Permeabilities are determined largely by the nature of the cell membrane as a hydrocarbon phase. Hence partitioning into this phase, well modeled by the partitioning behavior of organic solvents, is the major factor by which the membrane acts as a barrier to permeation. The nature of the hydrocarbon phase, a polymer-like organized lattice of hydrocarbon chains, creates the steep size dependence of permeability and

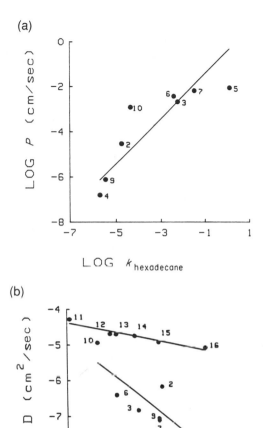

Fig. 2.9.   Permeability coefficients related to lipid solubility and to molecular mass. (a) Much as in Fig. 2.8, basal permeabilities (i.e., with all specific systems inhibited) of the human red blood cell membrane for eight different nonelectrolytes are plotted against their hexadecane partition coefficients. The permeants are numbered as follows: 2, ethanediol; 3, ethanol; 4, glycerol; 5, n-hexanol; 6, methanol; 7, n-propanol; 9, urea; 10, water. (b) Diffusion coefficients plotted against molecular mass. In the upper line the values are for diffusion in water; in the lower line the values are calculated for the hydrophobic phase of the human red cell membrane using Eq. (2.11) and measured values of permeability coefficients and hexadecane–water partition coefficients and a plausible value for the membrane thickness. The numbering of the solutes is as in (a) with also: 11, H_2; 12, N_2; 13, O_2; 14, CO_2; 15, urea; 16, glycerol. Taken from W. R. Lieb and W. D. Stein, in Stein (1986).

accounts for the sieving properties of the membrane. Molecules will cross cell membranes if they dissolve well in organic solvents and are small. They will cross poorly (unaided by specific systems of the cell membrane) if they are hydrophilic (polar) or large.

It is worthwhile to develop an intuitive feeling for the barrier properties of cell membranes, those that arise by virtue of their basic structure and that remain when we have inhibited the specific transport systems. Movement within the membrane occurs, as we have seen, by simple diffusion brought about by the ceaseless, random motion of the diffusing molecules. The net rate of movement from one side of the membrane to the other will depend on the number of such molecules diffusing. This number is higher if the permeant partitions readily into the membrane, lower if its solubility in the membrane phase is poor. A sugar molecule, for instance, is very poorly soluble in organic solvents. (Try dissolving sugar in olive oil!) Hence the number of sugar molecules available within a cell membrane and capable of diffusing across it will be low; its membrane permeability will, therefore, be low. Conversely, the anesthetic chloroform dissolves readily in cell membranes. Its concentration within the membrane will be high. Many molecules of chloroform will be present to diffuse, and its rate of permeation into cells will be extremely high. Polypeptides, nucleotides, intermediates of the tricarboxylic acid cycle, amino acids, and sugars (all compounds of profound importance for the living cell) are all very polar, and hence insoluble in organic solvents, and large in size. Unaided, they cross cell membranes very slowly, if at all, and the membrane acts as a very effective barrier keeping these important molecules within the cell after their synthesis. Oxygen and carbon dioxide are small molecules and highly soluble in organic solvents. They permeate cell membranes very effectively to be taken up or released during cell metabolism.

The data for artificial liposome membranes whose chemical composition can be controlled (such as those recorded in Table 2.3) show that the phospholipid composition of the membrane is an important determinant of membrane permeability. A higher percentage of *unsaturated* fatty acids renders the membrane more permeable to a broad range of solutes. One can, by diet, modify the composition of the membrane in bacterial cells and even in rats, with changes in membrane permeability that are in accordance with this rule. Increasing the temperature at which permeability measurements are made has much the same effect as increasing the unsaturated fatty acid composition of the membrane; permeabilities are increased across the board. The temperature coefficient is not the same for all solutes. Those solutes that permeate more slowly have in general higher temperature coefficients, i.e., they are more readily affected by an increase in temperature.

It is useful to keep in mind some rough measures of how chemical groupings on a permeant can be expected to affect its membrane permeability. Every hydroxyl group that is added to a molecule may be expected to *decrease* its permeability 100- or 1000-fold. A carboxyl group will have even a somewhat larger effect, while the addition of an amide residue is more or less equivalent to the addition of two hydroxyls. Conversely, every methyl group added is likely to *increase* permeability five-fold. A doubling of molecular volume will decrease permeability 30-fold. Adding charged groups, such as quaternary amines or carboxyl groups, decreases organic solvent partition coefficients and hence permeabilities by many orders of magnitude. Likewise, the common inorganic cations and anions have vanishingly low partition coefficients in such solvents and correspondingly low permeabilities across lipid membranes.

These considerations have been much used in the design of drugs required to be effective on targets *within* the cell. Adding methyl groups to a prototype drug increases its solubility in cell membranes and hence its ability to permeate into the cell. Benzyl rings are similarly effective and much used to improve the penetration of drugs. Conversely, drugs which are to act on targets *outside* the cell, i.e., on receptors facing the extracellular fluids of the cell, are designed so as to bear charged groups (for instance, choline residues or sulfate groups). These groups decrease enormously the solubility of the drug in question in organic solvents. Hence they greatly decrease membrane permeability and confine the access of the drug to the extracellular fluids of the body. These considerations have been used also in the design of slow-release drug-delivery systems in which drugs are incorporated into liposomes that are then injected into the body. A useful system is one in which the permeability of the drug across the membrane of the liposome is so low that its delivery to the target in the body is slow and controlled.

2.7. OSMOSIS AND THE DIFFUSION OF WATER

The point labeled "10" in Fig. 2.9(a) and (b) (see legend for Fig. 2.9) refers to the permeation of water across the membrane of the human red blood cell. Figure 2.9(a) shows that water crosses the membrane of this cell very rapidly, but the data plotted as the lower line of Fig. 2.9(b) show that this fact is by no means incompatible with the permeabilities of the other permeants studied. It would appear that water crosses this membrane rapidly simply because water is a very small molecule, the smallest of those depicted in Fig. 2.9, and has a partition coefficient that is not vanishingly small. Indeed, in most cells that have been studied, water crosses the membrane by the regular solubility–diffusion mechanism that is the basis of the simple diffusion that we studied in Section 2.6.

Note, however, that the data in Fig. 2.9 refer to the permeabilities that remain *after* any specific, parallel pathways in the red cell membrane have been blocked by appropriate inhibitors. Water can move on such a specialized pathway in the human red cell membrane, a pathway blocked by the addition of mercury-containing substances to the red cell (see the data collected in Table 2.6). In the presence of these mercurials, the rate of water movement across this membrane is greatly reduced. As we shall see in Section 7.2.2 (and as in the data collected in Table 2.6), in mammalian kidneys and in frog skin, the addition of antidiuretic hormones greatly increases the rate of water movement across the epithelia. Clearly, specialized pathways exist for water movement in these tissues, pathways of great importance for the effective functioning of the animal. (In mammalian kidneys, this rapid flow of water is essential for the ability of the kidney to retain body water in periods of restricted access to water, or in dehydration. Especially in desert-dwelling mammals, this pathway is of paramount importance to the organism.) How is this specialized flow of water carried out?

Before we consider this problem, we have to clear up another. Water is by far the major constituent of the solutions that exist within a body, or in which a body may find itself. The flow of this water brings about a net bulk flow of matter, a very visible volume change of the phase from which water is flowing or to which it moves. This visible volume change, often relatively easy to measure, has focused attention on the volume flow and led to the introduction of a different terminology in order to describe it. But the flow of volume is none other than the overall flow of those constituents of the solution that contribute to its volume. The phenomenon of bulk volume flow is known as *osmosis,* and we should now learn how the terminology of osmosis is related to the terminology of diffusion that we are familiar with from Eq. (2.7).

We repeat Eq. (2.7) here for convenience:

$$J_{I \to II} = PA(S_I - S_{II}) \qquad (2.7)$$

Consider a solution of, say, sucrose in water for the case where sucrose is totally unable to cross a membrane separating two solutions containing sucrose at two different concentrations. If the membrane is freely permeable to the aqueous component of the solution, the membrane is called, historically but misleadingly, *semipermeable* to sucrose. The flow of matter referred to on the left-hand side of Eq. (2.7) is, then, the flow of water. We can translate this into a bulk flow of volume if we realize that each molecule of water occupies a volume given by its partial molal volume— about 18 milliliters (ml) per mole. We multiply *both* sides of Eq. (2.7) by $\overline{V}$, the partial molal volume of water. The force that moves the flow of water is the difference in water concentration at the two faces of the membrane ($S_I - S_{II}$). But this difference of water concentration is merely the difference of concentration of the sucrose written with a change of sign, since solute and solvent together make up the whole solution on each side of the membrane. (For the case where sucrose$_I$ + water$_I$ = sucrose$_{II}$ + water$_{II}$, then water$_I$ − water$_{II}$ = sucrose$_{II}$ − sucrose$_I$.)

We can now obtain an equation that gives us the rate of volume flow of water by osmosis (written as $J^v_{I \to II} = \overline{V} J_{I \to II}$). This equation will be in terms of the osmotic permeability coefficient of the membrane (defined as $L_p = \overline{V} P/RT$) and the "osmolar concentrations" of the solutes at sides I and II of the membrane. These osmolar concentrations, Π_I and Π_{II}, are given by RT times the concentration (in conventional units) of *the solute particles of the nonpermeable matter.* [The number of solute particles is given by the number of molecules or ions of the solute, multiplied by the number of particles into which each dissociates in the solution in question, that is, by the "osmotic coefficient." The term RT comes from Eq. (2.3), which defines chemical potentials. Osmotic concentrations are conventionally in units of RT times molar concentration units.]

The desired equation [Eq. (2.13)] is obtained by multiplying both sides of Eq. (2.7) by $\overline{V}$ and making the appropriate substitutions:

$$J^v_{I \to II} = L_p A (\Pi_{II} - \Pi_I) \qquad (2.13)$$

Just like any solute, water will flow from one side of a membrane to the other when its concentration, or in other words, when the concentration of impermeable (nonwater) molecules, is different on the two sides. The flow of water is termed *osmosis*. The force that drives this volume flow, when considered as arising from the concentration difference of the impermeable solutes, is known as the *osmotic* pressure, and is given by RT times the concentration difference of nonpenetrating, and hence osmotically active, solutes at the two membrane faces.

We use the term *osmotic pressure* to remind us that water flow can also be brought about by applying a hydrostatic pressure difference across a membrane. Cell membranes can support very little pressure without breaking, but in bacterial and plant cells a rigid cell wall, external to the plasma membrane, gives this membrane mechanical support and allows the composite structure to withstand pressures of many atmospheres. In the animal body, the hydrostatic pressure developed by the action of the heart drives fluid across the capillary walls, between the cells that make up these tissues.

Mathematically, we can write an equation that relates the rate of fluid flow $J^v_{I \to II}$ to the difference in hydrostatic pressure (ΔP) across the membrane as

$$J^v_{I \to II} = L_p A (\Delta P) \qquad (2.14)$$

Consider, now, Fig. 2.10, which shows a membrane subject to a hydrostatic pressure difference and to an osmotic pressure difference. We apply a hydrostatic pressure across the membrane so as to bring about a flow of volume from, say, left to right. We set up a concentration difference of an impermeable solute so as to bring about an oppositely directed volume flow, from right to left. At a certain pressure difference the two volume flows will be not only oppositely directed, but equal. At this point, the hydrostatic pressure must be equal and opposite to the osmotic pressure. (See Box 2.4 for a discussion of this balance point in terms of the chemical potential of the water at the two sides of the membrane.) The same coefficient, which we have symbolized as L_p, can describe the relation between the force (the hydrostatic *or* the osmotic pressure difference) and the flow (that of volume) in the two cases, provided that the units that describe the terms in Eq. (2.13) and (2.14) are correctly chosen. Our concept of solute and solvent molecules impinging cease-

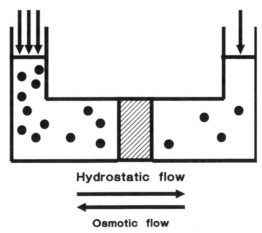

Hydrostatic flow

Osmotic flow

Fig. 2.10. Hydrostatic pressure and osmotic pressure oppositely directed across a membrane. The left-hand compartment contains a higher concentration of osmotically active molecules (and hence is at a higher osmotic pressure) than does the right-hand compartment. It is, however, subjected to a higher hydrostatic pressure, which *can* be arranged so as to balance the osmotic pressure, leading to zero net flow of the solute.

lessly on the surface of the cell membrane enables us to understand this. In the situation depicted in Fig. 2.10, there are simply more water molecules hitting that membrane surface at which the water is at the higher concentration. The momentum of these impacts is what drives the bulk flow of water from one membrane surface to the other.

Box 2.4. Osmotic Pressure and the Chemical Potential of Water

A helpful way of studying the relation between the osmotic pressure of a solution and the concentration of dissolved solutes is to return to Eq. (2.3), which contained the definition of chemical potential, i.e.

$$U_j = U_j^\circ + RT \ln S_j + \overline{V}_j P + Z_j F \psi \qquad (2.3)$$

It must be pointed out that the terms in this equation have to be defined differently for a solvent than for a solute. In the discussion concerning Eq. (2.3) (Section 2.1) we gave the definitions appropriate to a consideration of solutes, but we now want to discuss water and must use standard states and activities as defined for a solvent. For a solvent, the standard state is defined as the chemical potential of the pure solvent (and not of a 1 molal solution as for a solute). Instead of the concentration (or more accurately, the activity) we use the mole fraction of the solvent, i.e., the number of moles of the solvent as a fraction of the total number of moles present in the solution (solvent plus all the solutes). This gives a consistent set of definitions since in the pure solvent the mole fraction is unity, the logarithm of this is zero, and the chemical potential (at zero excess pressure and with no effect of electrical potential) is equal to the standard-state value. To check this, substitute the appropriate values in Eq. (2.3).

Now take the case where we have a setup similar to that of Fig. 2.10 but with pure water present in the right-hand compartment. Consider the equilibrium situation when there is no volume flow and a pressure difference, P, just balances the osmotic pressure difference, Π, owing to the presence of the dissolved solutes in the left-hand compartment. No water is flowing, hence the chemical potential of the water in the two compartments is equal. Substituting into Eq. (2.3) for the case of the right-hand compartment (pure water), and writing w in place of the subscript j, we have

$$U_w = U_w^\circ \qquad (B2.4.1)$$

(*Continued*)

the standard-state chemical potential of water, since the last three terms in Eq. (2.3) are equal to zero. For the case of the left-hand compartment we have (for Z_j equal to zero)

$$U_w = U_w^\circ + RT \ln a_w + \overline{V}_w P \qquad (B2.4.2)$$

where a_w is the activity of water in the left-hand compartment. Now the left-hand sides of these Eqs. (B2.4.1) and (B2.4.2) are equal, the chemical potential of water being equal in the two compartments. Also the standard state of water is the same in the two compartments. This means that at equilibrium

$$RT \ln a_w = -\overline{V}_w P \qquad (B2.4.3)$$

We can replace a_w by $1 - \Sigma(a_j)$, where the term with Σ refers to the sum of the mole fractions of all the solute components present in the left-hand compartment. When $\Sigma(a_j)$ is small, the logarithmic term becomes equal to $-\Sigma(a_j)$. Again, when $\Sigma(a_j)$ is small, this term itself becomes equal to the sum of the moles, n_j, of all the solutes present divided by the moles of water, $\Sigma(n_j)/(n_w)$. Equation (B2.4.3) then becomes

$$RT \Sigma(n_j) = (n_w)\overline{V}_w P \qquad (B2.4.4)$$

Finally, $(n_w)\overline{V}_w$ is equal to the volume of water present. Dividing both sides of Eq. (B2.4.4) by this term transforms the term on the left-hand side of Eq. (B2.4.4) into RT times the sum of the *concentrations* of the solute molecules present, $\Sigma(c_j)$, where we write c for concentration. This gives us the answer to the problem: the excess pressure required to prevent water flow from pure solvent into the solution, i.e., the osmotic pressure Π, is given by

$$P = \Pi = RT \Sigma(c_j) \qquad (B2.4.5)$$

One can see from this derivation that the result holds only at low concentrations of solutes, where the approximations made are realistic: that the logarithmic expansion is valid, that the moles of water present are approximately equal to the total moles present in the left-hand compartment, and that activities can be approximated by concentrations.

Volume flow is often measured by following the change in *light scattering* of a suspension of cells undergoing a volume change. As the cell shrinks or swells, the light scattered from each cell in the suspension alters and the suspension becomes more or less turbid, a phenomenon that can be followed optically. Figure 2.11 shows a typical experiment using light scattering. It depicts the time course of volume change when a red cell suspension originally in physiological saline is plunged into a solution containing urea (a solute that permeates across the red cell

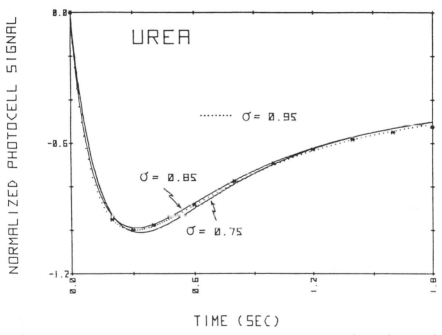

Fig. 2.11. Volume changes in a cell suspension as permeant enters and water leaves and then reenters. At zero time, human red blood cells were suspended in a concentrated solution of urea in physiological saline. The light scattered from the cell suspension is plotted as the solid stars on the ordinate against time. A downward deflection indicates cell shrinkage. The cell first shrinks as water rapidly leaves the cell in response to the osmotic gradient. Meanwhile, urea enters the cell and continues to enter. The entry of urea decreases the transmembrane osmotic gradient and water flows into the cell again, eventually restoring the original volume when the urea concentration is the same inside the cell as outside. The three theoretical curves are drawn with values of the reflection coefficient σ for urea of 0.75, 0.85, and 0.95. The last-mentioned gives the best fit to the data. (Reproduced from *J. Gen. Physiol.*, **81,** 239–253 by copyright permission of the Rockefeller University Press: Levitt and Mlekoday, 1983.)

membrane) in addition to the saline (which, effectively, does not). The cells first shrink as water leaves the cells, the osmotic pressure outside the cells being greater than that within them [Eq. (2.13)]. This part of the curve can be analyzed to provide a measure of the osmotic permeability coefficient for the red cell. Urea continually enters the cell, however, and adds to the osmotic material within the cell. Water flows down the ensuing osmotic gradient, and will continue to enter the cell, following the urea (we can call this an *osmotically obliged* flow of water), until at equilibrium, there is no difference in concentration of urea or of water on the two sides of the membrane.

2.8. DEFINITION OF REFLECTION COEFFICIENT

Clearly, the curve of Fig. 2.11 can be used to provide both the permeability of water (from the rate of *shrinking* of the cells) and the permeability of urea (from their rate of *swelling*). There is a problem, however, in the interpretation of such shrinking–swelling curves: if the permeabilities of water and the solute are sufficiently close together, a good deal of solute will have entered the cell before all the water has left osmotically. It becomes difficult (but not impossible) to dissect out the contributions of solvent and solute to the overall curve. Indeed, if the permeability of the solute is substantial it will not exert its full osmotic effect during the shrinking stage. This is a general problem that can bring us to a deeper understanding of osmotic relations. Equation (2.13) is correctly written only if the terms in it refer to the concentrations of a solute that is fully active osmotically. If a solute can enter the cell, the cell membrane is not wholly "semipermeable" for that solute. The osmotic effect of such a solute will be less than that predicted from the amount of it in the solution. It will be only a fraction of that predicted, somewhere between zero and unity according to whether the permeability of the solute is high or low.

We define this fraction as the *reflection coefficient,* generally symbolized by the Greek letter σ. Its value is the ratio of the osmotic pressure developed by a certain concentration of *that* solute compared with the osmotic pressure developed by the same concentration of a *fully impermeable* solute (i.e., σ = osmotic pressure of solute vs. osmotic pressure of fully impermeable solute). Obviously, σ is defined only in terms of a particular solute and a particular membrane. It is, in part, a measure of the membrane's permeability for that solute. A high permeability leads to a low value of the reflection coefficient, approaching zero. A low permeability will yield a reflection coefficient approaching unity. Table 2.5 records some reflection coefficients for various substances crossing different cell membranes.

TABLE 2.5

Reflection Coefficients for Various Compounds Crossing Membranes of a Range of Cell Types[a]

	Reflection coefficients			
Permeant	*Chara australis* (algal plant cell)	Human red blood cell	Urinary bladder (toad)	Squid axon (axolemma)
Formamide	1.0			0.44
Acetamide		0.58		
Malonamide		0.83		
Urea	1.0	0.62	0.79	0.70
Thiourea		0.85	0.995	
Ethanol	0.27			0.63
Ethanediol	0.27	0.63		0.72
Glycerol		0.88		0.96

	Kidney Tubules								
	Proximal convoluted		Descending thin		Cortical collecting		Papillary collecting		Distal convoluted	
	Rabbit	Rat	Rabbit	Rat	Rabbit	Rat	Rabbit	Rat	Rabbit	Rat
NaCl	0.68	0.68	0.96	<0.5	1.0	—	—	0.9	—	0.52
Urea	0.91	0.79	0.95	<0.5	1.0	—	—	0.4	—	0.67

[a] Mostly determined at 25°C. The data for *Chara*, red cells, toad bladder, and squid axon were determined by various authors and tabulated by W. D. Stein (1967) "The Movement of Molecules across Cell Membranes," Academic Press, New York. The data on kidney tubules are again from various authors and tabulated by R. L. Jamison and W. Kriz (1982) in "Urinary Concentrating Mechanism," Oxford University Press, Oxford.

A knowledge of the reflection coefficient of various solutes can be of great importance when we consider the movement of water in plants and animals. Thus, in the rat, sodium chloride and urea each have different reflection coefficients across the membranes that make up the different parts of the urinary tubule system of the kidney (see Section 7.2.1). In some regions, such as the collecting tubule, σ for salt is close to unity while that for urea is low. Thus, salt but not urea exerts almost its full osmotic effect. In other regions, such as the distal convoluted tubule, the reverse situation applies, the reflection coefficient for urea is high while that for salt is low. This situation is harnessed by the kidney in the mechanism whereby the urine becomes highly concentrated under conditions of extreme dehydration.

Interestingly, the value of σ can be obtained from measurements of the *minimum volume* in shrinking–swelling experiments such as depicted in

Fig. 2.11. If the reflection coefficient of the solute is unity (i.e., it is impermeable), the shrinkage will be maximal. If σ is zero, there will be no osmotic effect and no shrinkage. Any intermediate value of σ gives an intermediate volume change, and the minimum volume can be related, using the appropriate theory, directly to σ. From a single curve such as that of Fig. 2.11, L_p, σ, and the permeability of the solute can all three be derived. Such experiments are thus a rich source of data.

A final point that we must consider is the notion of the *interaction* between solute and solvent molecules. A solute crossing a cell membrane can bring with it, by the process of *solvent drag*, any water molecules that may be bound to it. When a permeable solute crosses the membrane in response to its concentration difference it may, by solvent drag, bring about a flow of water *in opposite direction* to the flow that is simultaneously being induced by the osmotic pressure difference (refer to Fig. 2.10). Thus solvent drag reduces the osmotic effectiveness of the solute. It has the effect, therefore, of further reducing the reflection coefficient, σ (defined a few paragraphs back). The deviation of σ from unity arises partly from the fact that the solute is permeable and hence cannot maintain an osmotic pressure difference at equilibrium, but also partly from the fact that a solvent drag can reduce net water flow during the approach to equilibrium. The flow of volume, the flow of solute and any interaction between solute and solvent are all three accounted for if the three parameters L_p, P_s, and σ are measured for the solvent–solute–membrane system in question.

2.9. COMPARISON OF OSMOTIC AND DIFFUSIVE FLOW OF WATER

Water is a very special molecule in that it can be regarded as both solute and solvent. Take a case where radioactively labeled water is added to side I of a membrane and its rate of flow is measured from I to II. This flow of label is not a bulk flow but is merely the diffusive flow of a particular chemical species, the labeled water. This flow is brought about by the concentration difference of the labeled water across the membrane. In no way can it be considered as being brought about by an osmotic pressure difference. It is simply the flow of water as solute. We can therefore define a coefficient P_w for water flowing as a solute (and measured isotopically). But, we can also define a coefficient L_p describing the bulk flow of water (measured as a volume flow) in the same system, under the same conditions. Do the two coefficients have to be identical? Both have been measured in many systems, and have often been found to differ significantly. Data for a few cases of the many systems that have been studied are collected in Table 2.6. For the human red blood cell, for instance, flow by osmosis is 5 to 10 times as fast as flow by diffusion.

TABLE 2.6

Osmotic Permeabilities (L_p) and Diffusive Permeabilities (P_w) Compared for a Range of Cells and Tissues and Artificial Systems[a]

System	L_p	P_w	Ratio	Ref.[c]
Artificial lipid bilayers	20–50	1–11, but under-estimated	Probably unity	(i)
Bilayer with gramicidin channel added	9.6[b]	1.8[b]	5.3	(ii)
Red blood cells: a	200	20–40	5–10	(iii)
Inhibited by mercurials: b	12–18	12–18	1	(iii)
Through channels: a − b	185	10–20	9–18	(iii)
Proximal tubule cells, rabbit kidney				
Uninhibited	400	22	18	(iv)
Inhibited by mercurials	32	10	3.2	(iv)
Toad urinary bladder				
Unstimulated: a	2–4	0.7	3–6	(v)
Stimulated with antidiuretic hormone: b	41	1.9	22	(v)
Through channels: a − b	39	2	16–20	(v)

[a] Values expressed as 10^4 cm sec^{-1}, except for gramicidin channel.

[b] Units here are per channel, in cm^3 sec^{-1}.

[c] Key to References: (i) R. Fettiplace and D. A. Haydon, *Physiol. Rev.* **60**, 510–550 (1980); (ii) A. Finkelstein, *Curr Topics Membr Transp.* **21**, 295–308 (1984); (iii) W. D. Stein, (1984) "Transport and Diffusion across Cell Membranes," pp. 150–151 from various authors. Academic Press, Orlando, Florida; (iv) P. Carpi-Medina, V. León, J. Espidel, and G. Whittembury, *J. Membr. Biol.* **104**, 35–43, (1988); and (v) S. D. Levine, M. Jacoby, and A. Finkelstein, *J. Gen. Physiol.* **83**, 543–561 (1984).

Clearly such a finding must throw light on the mechanism by which water flow is occurring across any membrane for which such a difference is found. Take the case of water flowing through a channel or pore as depicted in Fig. 2.12. Flow by osmosis is brought about by the osmotic pressure difference. The pressure applied on one end of the long chain of water molecules that extends through the pore is mechanically transmitted from one side to the other. When a water molecule is forced, by the applied pressure, to enter the channel from one side of the membrane, another water molecule is ejected at the opposite face and a net flow of water takes place. The rate at which flow is occurring is determined exactly by the pressure difference between the ends. Thus Eq. (2.13) applies, and the whole of the osmotic pressure difference will affect flow. Not so for the concentration difference! For a labeled water molecule

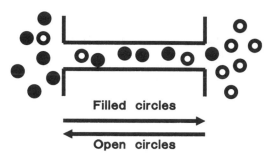

Fig. 2.12.   Flow of water through a long, narrow pore. Water molecules are depicted flowing through a long, narrow pore. The molecules entering the system from the left-hand side are drawn as the filled circles, those from the right-hand side as empty circles. Within the pore the molecules cannot pass one another.

(say, one of the empty circles in Fig. 2.12) to leave at the left, after first entering at the right, it must exchange places with *all* of the water molecules ahead of it in the chain. Flow of label will be greatly reduced by this ''long channel'' or ''no-pass'' effect and a great difference will be found between the measured values of L_p and of P_w. Indeed, it can be shown that the ratio of these coefficients is approximately equal to the *number* of water molecules that lie in a chain across the membrane. This enables us to estimate the length of the pore across the membrane. For the human red blood cell, where the osmotic:diffusive flows ratio is about ten, it would seem that the channels through which water flows can hold nine or ten water molecules in line across the membrane. For the proximal tubule cells of the kidney, this number may be as high as 17. We will consider other such measurements in Sections 3.1 and 3.2. An additional method of estimating the length of a channel is described in Box 2.5.

It is important to realize that, for a permeant which crosses the membrane as rapidly as does water, the presence of the unstirred layers discussed in Section 2.4 (Box 2.3) has to be taken into account. The data recorded in Table 2.6 are from studies where this problem has been satisfactorily overcome.

Note that the movement of water across cell membranes, when no specific channels are operating, can be described by a permeability coefficient of approximately $10–20 \times 10^{-4}$ cm^2 sec^{-1}. This is about the rate at which water crosses the artificial lipid bilayers listed in Table 2.6. There is nothing to suggest that water moves across unmodified cell membranes by any other route than the solubility–diffusion mechanism that we discussed in Section 2.6. On the other hand, water appears to move through the specialized systems (those in Table 2.6 inhibited by mercurials or stimulated by hormone) as if it traveled through a pore or channel. In

---

Box 2.5. Estimating the Length of a Channel Using Electroosmosis and Streaming Potential Measurements

If a channel is narrow enough, it can happen that an ion moving through the channel will sweep water molecules ahead of it. This gives rise to the phenomenon of *electroosmosis,* the flow of water induced by the passage of an electric current. The reverse of this phenomenon is the *streaming potential,* arising from the flow of ions that may be carried through a narrow channel when water is forced osmotically to flow through it. The osmotic flow of water may give rise to an electrical potential that can, under favorable circumstances, be measured across the membrane. The two phenomena, the streaming potential and electroosmosis, arise from *coupling* between the mutual flows of water and of ions. It can be shown [see Stein (1986), pp. 155, 156; or any physical chemistry textbook] that the number of water molecules moving through the channel for every ion that is flowing is given by n in the expression

$$n - (F/RT\overline{V}_w)(-\Delta\psi/\Delta P) \qquad (B2.5.1)$$

where F is the Faraday constant, R the gas constant, T the absolute temperature, and $\overline{V}_w$ is the partial molar volume of water (18 cm³ mol⁻¹). $\Delta\psi$ (in volts) and ΔP (in osmolars) are the measured voltage and applied osmotic pressure in a streaming potential measurement, or the applied transmembrane voltage and the resultant osmotic pressure developed in electroosmosis. Substituting appropriate values for the fundamental constants (Appendix 2) into Eq. (B2.5.1) reveals an easy-to-remember relation: An osmotic pressure difference of 1 osmolar will give rise to a transmembrane potential of 0.45 mV for each water molecule in the no-pass channel.

Chapter 3 we go on to discuss the membrane channels through which water often moves and through which, in many cases, specific ions can travel.

SUGGESTED READINGS

General

Stein, W. D. (1986). "Transport and Diffusion across Cell Membranes," Chapters 1 and 2. Academic Press, Orlando, Florida.

Diffusion as a Random Walk

Einstein, A. (1926). "Investigations on the Theory of the Brownian Movement" (R. Furth, ed.; A. D. Cowper, transl.). Methuen, London.

Chemical Potential

Nobel, P. S. (1974). "Introduction to Biophysical Plant Physiology." Freeman, San Francisco, California.

Electrical Potential

Hille, B. (1984). "Ionic Channels of Excitable Membranes." Sinauer, Sunderland, Massachusetts.
Blaustein, M. P., and Lieberman, M. (1984). "Electrogenic Transport Fundamental Principles and Physiological Implications." Raven Press, New York.

Flux Ratio Test

Ussing, H. H. (1949). Distinction by means of tracers between active transport and diffusion. The transfer of iodide across isolated frog skin. *Acta Physiol. Scand.* **19**, 43–56.

Permeability and Partition Coefficients

Walter, A., and Gutknecht, J. (1986). Permeability of small nonelectrolytes through lipid bilayer membranes. *J. Membr. Biol.* **90**, 207–217.

Measurement of Permeability Coefficients

Brahm, J. (1983). Permeability of human red cells to a homologous series of aliphatic alcohols. *J. Gen. Physiol.* **81**, 283–304.
Eidelman, O., and Cabantchik, Z. I. (1989). Continuous monitoring of transport by fluorescence on cells and vesicles. *Biochim. Biophys. Acta.* **988**, 319–334.
Levitt, D. G., and Mlekoday, H. J. (1983). Reflection coefficient and permeability of urea and ethylene glycol in the human red cell membrane. *J. Gen. Physiol.* **81**, 239–253.

NMR and ESR

Watts, A. (1989). Membrane structure and dynamics. *Curr. Opin. Cell Biol.* **1**, 691–700.

Unstirred Layers

Barry, P. H., and Diamond, J. M. (1984). Effects of unstirred layers on membrane phenomena. *Physiol. Rev.* **64**, 763–872.

Plant Cell Permeabilities

Collander, R. (1954). The permeability of *Nitella* cells to nonelectrolytes. *Physiol. Plant.* **7**, 420–445.

Membrane as a Hydrophobic Sieve

Lieb, W. R., and Stein, W. D. (1986). Non-Stokesian nature of transverse diffusion within human red cell membranes. *J. Membr. Biol.* **92,** 111–119.

Osmosis and the Diffusion of Water

Finkelstein, A. (1986). "Water Movement through Lipid Bilayers, Pores and Plasma Membranes." Wiley, New York.

Reflection Coefficient

Katchalsky, A., and Curran, P. F. (1965). "Nonequilibrium Thermodynamics in Biophysics." Harvard Univ. Press, Cambridge, Massachusetts.

Electroosmosis and Streaming Potential

Levitt, D. G. (1984). Kinetics of movement in narrow channels. *In* "Ion Channels: Molecular and Physiological Aspects" (W. D. Stein, ed.), Curr. Top. Membr. Transp., Vol. 21, pp. 181–197. Academic Press, Orlando, Florida.

Ion Channels across Cell Membranes

The solubility of most ions in organic solvents is negligible; hence the concentration of an ion in the hydrophobic barrier that constitutes the cell membrane is vanishingly small. The physical reason for this low solubility is the enormous energy that would have to be invested to transfer a charged particle from a high dielectric medium such as water (dielectric constant of about 80) to a low dielectric medium such as an organic solvent or a cell membrane interior (dielectric constants of 2–4). (For the meaning and use of dielectric constants, a textbook on physics should be consulted.) This energy is about 80 kcal/mol (335 kJ/mol), compared with an energy of 4–5 kcal/mol (21 kJ/mol) required to transfer a single hydroxyl group from water to an organic phase. (See Box 3.1 for a fuller treatment.)

Were ions to cross cell membranes only by the solubility–diffusion mechanism used by nonelectrolytes (as described in Section 2.3), this rate would be almost immeasurable. Yet there is a constant traffic of ions across all cell membranes. Sodium, potassium, calcium, protons, chloride, and bicarbonate ions (and many others) are all needed by cells and pass in and out of cells rapidly, in a controlled fashion, well adjusted to cellular requirements.

How is this flow brought about? It has been found that *rapid* ion flow occurs through channels, which behave in principle like that depicted in Fig. 2.12 of the previous chapter. In many cases, the protein molecules which constitute the ion-specific channels have been isolated and purified from the cell membranes in which they are present. In certain cases, the genes coding for the proteins that constitute the ion channels have been

Box 3.1. Calculations of the Electrostatic (Born) Free Energy

Consider a sphere of radius r and composed of a substance of dielectric constant ε_s, situated in a medium of dielectric constant ε_m. From classical electrostatic theory (as reviewed in any physics textbook), the electrostatic energy required to put a charge q at the center of this sphere is $E = (16.6\ q^2/r)(1/\varepsilon_m - 1/\varepsilon_s)$, in units of kcal/mol when r is in nanometers. The work required to move this charge from a medium of dielectric constant ε_1 to a medium of dielectric constant ε_2 is thus given by $(16.6\ q^2/r)(1/\varepsilon_2 - 1/\varepsilon_1)$, the partition coefficient, by $\exp(-\text{work}/RT)$. Substituting 1 for the charge q, the reasonable value of 0.1 nm for the radius, and 80 and 2 for the dielectric constants of water and the hydrophobic phase that is the membrane, respectively, gives a computed energy of 81 kcal/mol (339 kJ/mol).

The partition coefficient for such an ion will therefore be given by $K = \exp(-81/RT)$. With RT at 0.594 kcal/mol at 27°C, the partition coefficient is 6×10^{-60}, an immeasurably small amount. (In contrast, a free energy of 4 kcal/mol for the partitioning of an hydroxyl group between water and lipid gives a partition coefficient of 1.2×10^{-3}, an easily measurable value.) In the example just treated, the ion had a valence of 1. If, rather, a divalent ion is considered, the free energy is increased fourfold and the partition coefficient falls to less than 10^{-114}.

cloned (Boxes 3.3 and 3.7, later in this chapter) and their amino acid sequences determined. The channel molecules are being intensively studied by physical techniques such as electron microscopy and X-ray diffraction, giving us a detailed picture of their structure. In quite a few cases, we have been able to reach an informed understanding of ion movements across channel-containing membranes. We shall discuss a few types of ion channels in some detail, in Sections 3.1 and 3.2, and again in Section 3.6.

3.1. THE GRAMICIDIN CHANNEL

The bacterium *Bacillus brevis* produces a polypeptide called *gramicidin,* an important antibiotic that effectively kills gram-positive bacteria. Analysis of the mechanism of action of gramicidin showed that it causes a loss of ions from those bacteria against which it is effective. Later, it was shown that adding a small amount of gramicidin to a cell membrane or to a lipid bilayer (see Section 1.3) greatly increases the rate at which cations pass across the "doped" membrane. Gramicidin acts as an *ionophore*, that is, a substance that one can add to a lipid bilayer and thereby increase greatly the rate at which ions move across it. (In Chapter 4, we shall come across a second class of ionophores that act in another way, namely, as carriers of ions rather than as channels). Figure 3.1(a) depicts the chemical structure of the gramicidin molecule, a 15-membered chain of alternating D- and L-amino acids. Figure 3.1(b) shows a model, based on physical analyses, for the three-dimensional form of the gramicidin dimer that inserts into cell membranes. Note the hole running right through the molecule. Gramicidin is clearly a channel and indeed has served as an excellent model for the behavior of membrane channels.

The rate of ion movement across lipid bilayer membranes is generally measured electrically, by *conductance measurements* that measure the strength of the current that the ions carry across the membrane when an electrical potential is applied across it [Fig. 3.2(a)]. Conductance, which is the reciprocal of resistance, is defined as current passing divided by the electrical potential difference that drives the current flow. If current is measured in amperes and potential in volts, their ratio, conductance, is given in siemens.

In the case of gramicidin, the current is carried by the cations, since anions have an insignificant permeability through this molecule. When the concentration of gramicidin added is low, the time course of flow of current across the membrane shows the characteristic appearance depicted in Fig. 3.2(b), a series of fluctuations in current, in which the current at any instant is at one of a series of discrete values. The step from one value of the current to another is extremely rapid. Dennis Haydon and Stephen Hladky saw the importance of this. They realized that these fluctuations arose from the discrete insertion of gramicidin (as dimers) into the membrane. At the very lowest current flow, there would be *only a single molecule* of gramicidin present in the membrane, while at higher levels two, three, or [in Fig. 3.2(b)] up to five molecules of gramicidin were present simultaneously. The lowest step gives the current carried across the membrane through a single dimer of gramicidin. This is termed the *single-channel conductance* of the gramicidin channel. It can be calculated that under the conditions illustrated in Fig. 3.2, a single gramicidin molecule can allow 1.5×10^7 ions to pass through it per second! A conveniently sized unit to report single-channel conductances is the picosiemen (pS, 10^{-12} siemens). Measured between two 100 mM CsCl solu-

(a)

Formyl-L-Val-Gly-L-Ala-D-Leu-L-Ala-D-Val-L-Val-D-Val-
      1   2   3   4   5   6   7   8

L-Trp-D-Leu-L-Trp-D-Leu-L-Trp-D-Leu-L-Trp-NHCH$_2$CH$_2$OH
  9   10   11   12   13   14   15

(b)

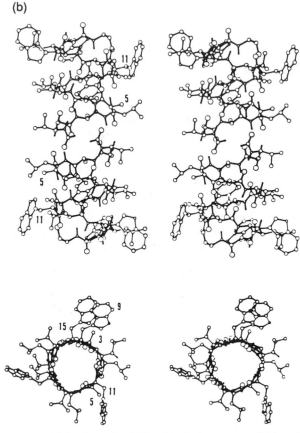

Fig. 3.1. The gramicidin channel. (a) The chemical structure of the molecule, an alternating L/D polypeptide. (b) A model for the structure of the channel, based on physical measurements. These are stereo projections and can be viewed in three dimensions by squinting at the pictures. The upper half of the figure shows a side view, the lower half an end-on view through the hole down the axis of the molecule. Taken, with kind permission, from Venkatchalam and Urry *J. Comput. Chem.* **4,** 461–469 (1983).

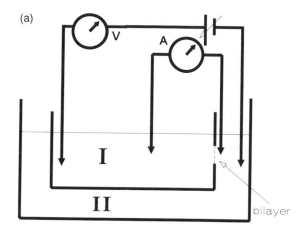

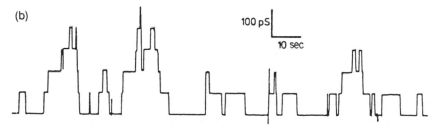

Fig. 3.2. Single-channel conductance measurements on gramicidin. (a) Schematic picture of the experimental setup. Sections I and II are two aqueous compartments separated by a small hole across which a bilayer of phospholipid molecules is spread. A small amount of gramicidin is added to the bathing solutions and dissolves in the bilayer. An electrical potential can be applied across this bilayer with the voltage source V and the resulting current flowing across it measured by the ammeter A. (b) An example of the current flowing across such a bilayer, as a function of time. The current rises and falls in a series of steps, all of them multiples of a unit current flow. From a knowledge of the applied potential and this current flow, the membrane's conductance can be calculated. Taken, with kind permission, from Hladky and Haydon (1984).

tions, the single-channel conductance of the gramicidin channel is about 30 pS. (Check a physics text to confirm that a single-channel conductance of one picosiemen is equivalent to a flow of 6.28×10^6 ions transported through the channel per second per applied volt.)

If ions can flow through the gramicidin molecule, it is not surprising that water molecules, too, can flow through it. When, indeed, the rate of water movement through the gramicidin channel was measured by the two methods described in Section 2.9—by measurement of the flow of isotopically labeled water, on one hand, or by the osmotically induced bulk flow of water, on the other—it turned out that bulk flow was substantially *faster* than isotopic flow. This recalls the discussion concerning Fig. 2.12. The data have been interpreted in terms of *interference* between water molecules in a narrow, no-pass channel extending through the gramicidin molecule and, indeed, provide very good support for the suggestion that gramicidin *is* a true channel. From the ratio of bulk to isotopic flows, the biophysicists David Levitt and Alan Finkelstein have independently calculated that approximately six to nine water molecules form a chain across the cell membrane through the gramicidin molecule. The length of this chain fits very well with the estimates of the dimensions of the gramicidin molecule as deduced from physical measurements (Fig. 3.1). The fact that water molecules interfere with one another's passage across the membrane suggests that the gramicidin channel is narrow, smaller in cross section than two water molecules. Measurements of streaming potentials (Box 2.5 in Chapter 2) confirm this conjecture, giving 6–10 as the number of water molecules that lie along the length of the gramicidin channel.

When the conductance of a gramicidin-doped membrane is measured for a number of different cations as a function of cation concentration, results such as are depicted in Fig. 3.3 are obtained. Clearly, the rate of flow of ions through the channels reaches, or goes through, a maximum as the concentration of ions in the bulk solution is increased. We say that the flow of ions appears to *saturate* as the ion concentration is increased. Why should a maximum be reached? It must mean that the ions themselves must interfere with one another as they attempt to flow through the channel. Indeed, the most likely explanation of Fig. 3.3 is that the entry of an ion into the gramicidin channel effectively prevents the entry into this channel of any other ions. (For a somewhat fuller treatment of such saturation curves, see Box 3.2.)

Presumably, the electric charge that the ion brings with it into the channel is sufficiently great to shut out other ions of like charge. The channel is seldom more than singly occupied up to the maximum in the conductance plot. At higher concentrations, however, a second ion appears to be able to enter the channel even against the electrostatic force of the first and, in so doing, each ion blocks the movement of the other ion in the channel, giving the reductions in conductance seen in Fig. 3.3. Movement through the gramicidin channel is thus consistent with a model in which one or (at most) two ions are present in the channel at any time, entering, diffusing through, and leaving it again by the random, statistical motions characteristic of all diffusion, as was seen in Chapter 2.

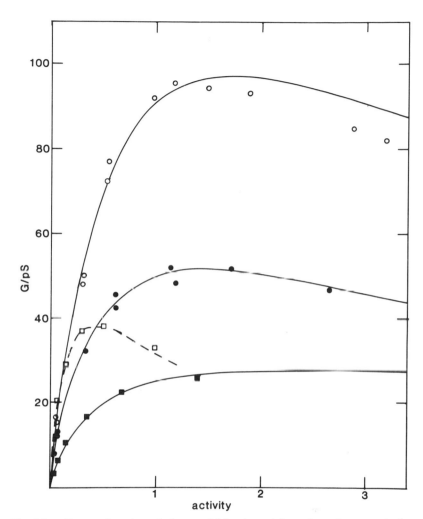

Fig. 3.3. Current flow through the gramicidin channel depends on the concentration of the permeating ions. The ordinate plots the conductance [in picosiemens (pS)] of a bilayer doped with gramicidin as a function of the activity (in molals) of the cation that carries the current through the channel. The four curves are for (from uppermost to lowest): CsCl (○), TlCl (□, dashed), KCl (●) and NaCl (■). Data measured at 21°C with an applied transmembrane potential of 50 mV. Taken, with kind permission, from Hladky and Haydon (1984).

Box 3.2. Saturation of Flows as Concentration Increases

Data such as the lowest curve in Fig. 3.3, in which the rate of flow of the ion increases smoothly with its concentration until a plateau is reached, are characteristic of a vast number of transport systems and enzymatic reactions. The curve is said to show saturation behavior. The rate of flow reaches a maximum termed the *maximum velocity* symbolized by V_{max}. One-half of this maximum velocity is reached at a characteristic concentration of the permeant (or substrate in the case of an enzyme), termed the *Michaelis parameter* (K_m), named for a pioneer of enzyme kinetics, Leonor Michaelis. The mathematical form of the equation that relates v, the rate of transport (or of the enzymatic reaction), to $[S]$, the concentration, is

$$v = \frac{[S]V_{max}}{[S] + K_m} \qquad (B3.2.1)$$

This equation is called the Michaelis–Menten equation (M. L. Menten was a coauthor of the 1913 paper that defined the parameters of this equation.)

The reader might care to check that, as $[S]$ tends to infinity, v tends to V_{max} while, when $[S] = K_m$, v is exactly one-half of V_{max}, consistent with the definition of these terms in the preceding paragraph. An extended discussion of this Eq. (B3.2.1) and its physical meaning will be found in Sections 4.2 and 4.6.

3.2. THE END-PLATE (ACETYLCHOLINE RECEPTOR) CHANNEL

At the place at which a nerve innervates a muscle or at the transmission gap between one nerve and another [that is, the nerve–muscle or the nerve–nerve synapse (see any physiology textbook)], channel-forming molecules are found embedded in the postsynaptic membrane, that beyond the gap. One such molecule is the *end-plate channel* or *acetylcho-*

line receptor. This binds the neural transmitter acetylcholine and, on so doing, rapidly undergoes a change that allows it effectively to conduct a variety of ions. By so doing, it reduces the transmembrane potential, that is, it "depolarizes" the postsynaptic cell membrane. We call this control of the opening of the channel (such as acetylcholine brings about) "gating." The acetylcholine receptor has been isolated from cell membranes. It has been shown to be a protein, its polypeptide composition has been determined, and the polypeptides have been sequenced. The availability of partial sequences enabled Shosaku Numa and colleagues to clone the genes that code for the polypeptides that form the acetylcholine receptor (see Box 3.3).

The acetylcholine receptor contains five polypeptide subunits (two being identical), each containing four membrane-spanning stretches (see Chapter 1, Section 1.4). The whole structure extends about 10 nm across the membrane, and has a cross-sectional diameter of some 9 nm [see Fig. 3.4(b) for a physical image of the receptor and Fig. 3.4(c) for a model]. Most interestingly, two other receptor channels [one gated by glycine, the other by γ-aminobutyric acid (GABA), both being anion-selective] have also been cloned and sequenced, yielding results remarkably similar to those obtained for the end-plate channel, showing a similar pattern of four polypeptide chains each containing four putative membrane-spanning stretches [Fig. 3.4(d)]. It seems that there is a family, probably related by common descent, of these chemically gated channels. The glycine and GABA receptors are more closely related to each other than either is to the acetylcholine receptor.

The end-plate channel is of low specificity. It allows many cations to pass through it, but is impermeable to anions. Small nonelectrolytes cross it with permeabilities that are one-tenth that of cations of like size. The channel saturates as does the gramicidin channel discussed above. Estimates of the cross-sectional radius of the pore have been made. Consider Fig. 3.5, which shows how the permeability of many different cations (simple inorganic cations as well as more complex organic cations) across the end-plate channel is related to the molecular weight of the penetrating ion. The line drawn through the points describes the behavior of a channel with a cross-sectional diameter of 0.74 nm, taking into account friction between the ion and the walls of the channel. Clearly a hole of this size provides a good model for the size discrimination that the acetylcholine receptor channel displays. Measurements of the streaming potential (see Chapter 2, Box 2.5) suggest that about six water molecules can extend *along the length* of this narrowest region of the channel.

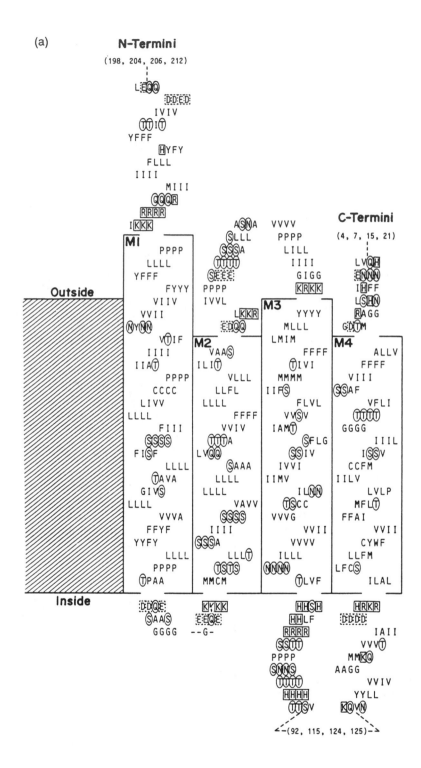

(a)

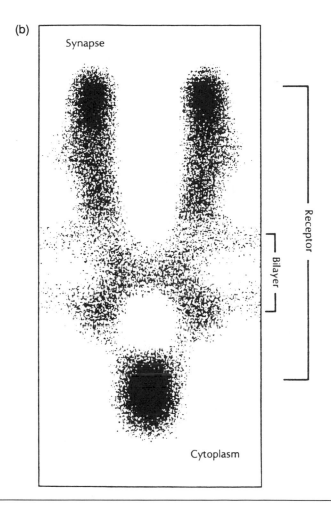

Fig. 3.4. Structure of the end-plate channel (the acetylcholine receptor). (a) Transmembrane disposition of all four polypeptide chains. Charged residues in boxes; hydroxyls encircled. Taken, with kind permission from S. Numa *et al.* (1983) *in* Cold Spring Harbor Symposia on Quantitative Biology, Vol. 48, Molecular Neurobiology. (b) A section through the end-plate receptor at 1.7 nm resolution, as deduced from electron microscope studies using image reconstruction. The position of the bilayer is shown, together with details of the funnel-shaped channel. Reprinted by permission from Toyoshima and Unwin (1988) *Nature* **336,** 247–250, copyright © Macmillan Magazines Ltd. (c) A model for how the five subunits are arranged to form the channel. The diagram shows a cross section through the molecule. Taken, with kind permission, from J. A. Dani (1989) *Curr. Opin. Cell Biol.* **1,** 753–764. (d) A model for the GABA$_A$ (γ-aminobutyric acid) receptor as deduced from the sequences of the α and β subunits that compose it. Membrane-spanning helices are depicted as cylinders traversing the phospholipid bilayer. Reprinted by permission from Schofield *et al.* (1987) *Nature* **328,** 221–227, copyright © Macmillan Magazines Ltd. (*Figure continues*)

(c)

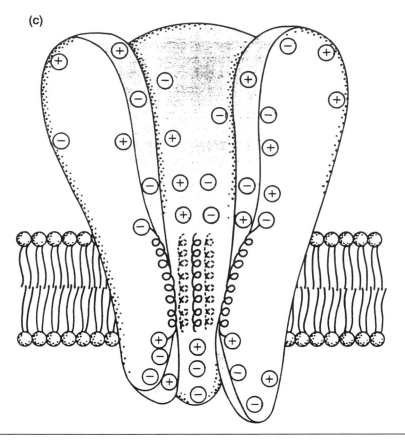

Fig. 3.4   (*Continued*)

Box 3.3. Molecular Biology of the End-Plate Channel

The end-plate channel or acetylcholine receptor channel was the first channel molecule to have the genes that code for it sequenced. We will describe briefly the methods used in this work by Numa and colleagues, since similar techniques are the basis for much of the molecular biology studies described in later chapters of this book. (For a more detailed description of the methods used in molecular biology and genetic engineering, a cell biology textbook should provide the needed background.)

(Continued)

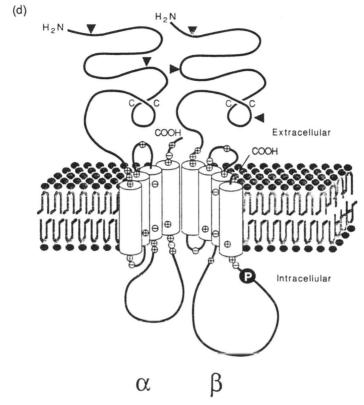

Fig. 3.4   *(Continued)*

The first step was to prepare the messenger RNA molecules from the electric organ of the fish *Torpedo,* as this organ is a rich source of the receptor and hence of the messenger RNA molecules that code for its polypeptides. The preparation contained, of course, the messenger RNA molecules coding for *all* the proteins of the organ, not just those that constitute the receptor. From this RNA, using the enzyme reverse transcriptase, they prepared the DNA complementary to the RNA of the organ, cDNA (complementary deoxyribonucleic acid). Individual DNA molecules from this preparation were inserted into the DNA of plasmids (minichromosomes) that were used to transform *Escherichia coli* cells. From these cells, *clones* were grown, each clone containing at random a molecule of the cDNA from electric organ. Next, thou-

(Continued)

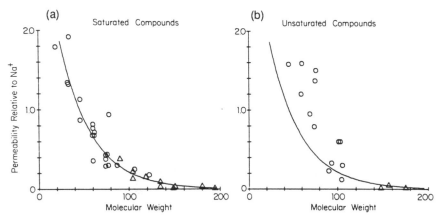

Fig. 3.5. Permeability of the end-plate channel depends on the size of the permeating molecule. Permeability (relative to that of sodium ions) through the end-plate channel, measured as described in Box 3.5, is plotted as a function of the molecular mass of the permeating cation. The data are recorded for saturated cations in (a); in (b), for unsaturated species. The line in (a) is drawn through the data and is reproduced as such in (b). Reproduced from T. M. Dwyer, D. J. Adams, and B. Hille (1980) *J. Gen. Physiol.* **75**, 469–492 by copyright permission of The Rockefeller Press.

sands of these clones were screened by attempts to hybridize the DNA *they* produced with DNA fragments synthesized so as to code possible polypeptide sequences of the receptor. The clones that did, indeed, contain DNA coding for the receptor sequences were isolated and, from these, sufficient DNA was prepared to allow the relevant genes to be *sequenced*. Four genes were found to code for the four different polypeptides that constitute the receptor. Later, similar genes were located in the genome of the cow and of the human. The transmembrane disposition as deduced from the sequences of the genes that code for the subunits of the end-plate channel is shown in Fig. 3.4(a).

In a subsequent study from Numa's laboratory, each cDNA coding for the four polypeptides was joined to a DNA sequence that is present in the simian virus SV40. This virus infects monkey cells and this particular DNA sequence (the promoter) enables DNA attached to it to be expressed to high levels in monkey cells, producing the appropriate messenger RNA in high yields. These messenger RNA molecules were isolated from the monkey cells and introduced into an *expression system* consisting of the oocytes (unfertilized egg cells) of the frog *Xenopus laevis,* in which the messenger RNA is translated into the corresponding polypeptide. In the case of polypeptides that bear a

(Continued)

sequence directing their synthesis into the membrane (Section 1.5), the polypeptide can be inserted correctly into the membrane of the oocyte. The four polypeptides could be expressed separately or in any combination in the oocyte, and the resulting effect on function was tested. The functions studied were the ability of the polypeptides to insert correctly into the membrane, the ability of the expressed polypeptides to bind drugs that inhibit the receptor and, finally, electrical effects. When the messenger RNA coding for all four polypeptides of the receptor was expressed in the oocyte, the receptor was fully functional and the addition of acetylcholine resulted in a flow of current across the membrane of the oocyte.

Numa and colleagues were also able to introduce point mutations into the DNA molecules coding for the four polypeptide chains, at specific sites in the sequence. In particular, they mutated at the amino acids that appeared to form three rings of negatively charged residues along the length of the channel (see also Box 3.4). Using a DNA synthesizer, they prepared stretches of DNA corresponding to these sequences and substituted them for the wild-type DNA. This is *site-directed mutagenesis*. These mutated DNA molecules were then cloned in a plasmid system as before. Messenger RNA corresponding to the mutated DNA molecules was synthesized and then expressed in the oocyte system. Patch-clamp studies (see Section 3.4.1) enabled the conductance of single channels formed by these mutated receptors to be measured. Thus the role of individual amino acid residues of the polypeptides that constitute the receptor could be studied. Some results of these experiments are depicted in Box 3.4.

3.3. CONDUCTANCES AND CROSS-SECTIONAL AREAS OF SINGLE CHANNELS

The simplest model one can think of for an ion-carrying channel is a cylindrical pore of uniform cross section extending right across the cell membrane and selecting between permeants merely by virtue of the size of the aperture of the pore. This model gives a surprisingly good account of the behavior of certain channels, as we shall show. Such a channel will not, however, distinguish between ions of different charge, i.e., between anions and cations. But we know that most membrane channels are exceedingly good at making this distinction. (Cation-selective channels are most common; anion-selective channels are found less frequently.)

If one assumes that within the mouth of most channels is a region of net negative charge (contributed by the amino acid side chains that.form the channel), this region will attract cations, raising their concentration above that of the bulk solution, and hence increasing the number of impacts that these cations make with the channel aperture (see Box 3.4). Just as we saw for the transmembrane movement of nonelectrolytes, any such *partitioning* effect will increase permeability. The ability of the channel to pass ions of charge opposite to that of a charge placed at the mouth (a so-called fixed charge) is increased while its ability to pass ions of like charge to the fixed charge is decreased.

Having accounted for the charge selectivity of the ion channels, one needs next to account for selectivity *among* ions of *like* charge. Let us test the assumption that this is, to a first approximation, a function of the size of the channel's opening. To make this test we can use Eq. (2.2), which we reproduce here:

$$J_{I \rightarrow II} = DA(S_I - S_{II})/d \qquad\qquad (2.2)$$

This gives the flow of material (here the ion in question) through a pore system of total cross-sectional area A and effective length d, driven by a concentration difference $S_I - S_{II}$, for a substance having the diffusion coefficient D. We can apply this equation just as well to a *single channel* (where the channel in question is far from being fully occupied by the penetrating ion), when A becomes the cross-sectional area of the single channel and d its length. (Some typical numbers are listed later in Fig. 3.7.) If the channel is occupied by water, we might hazard, as a first approximation, that the diffusion coefficient of the ion in *water* might be appropriate for diffusion through the channel.

Physical chemistry textbooks show that with the parameters Z, F, R, and T defined in Section 2.1, the conductance of a 1-cm cube of a solution containing a monovalent ion at concentration S_I, with diffusion coefficient D, is given by

$$\text{Conductance} = DS_I ZF^2/RT \qquad\qquad (3.1)$$

For a channel with cross-sectional area A and length d, the conductance is simply (A/d) times the value calculated using Eq. (3.1). Guessing that the diffusion coefficient of an ion in water is appropriate for an ion diffusing through an ion channel, we can use this expression to calculate conductances, all other terms being fundamental constants or quantities for which we can make reasonable estimates.

Table 3.1 records experimentally determined values of single-channel conductances (in picosiemens per molar) for six different channels whose cross-sectional areas have been estimated, together with values for the single-channel conductances calculated on the model that the ion flow is

TABLE 3.1

Calculated and Measured Equivalent Conductances for Some Single Channels for Which Estimates of Cross-Sectional Channel Area Are Available[a]

Channel	Estimated cross-sectional area ($\mathring{A}^2$)	Equivalent Conductance (pS M^{-1})	
		Calculated	Measured
Frog muscle end-plate (sodium)	40.3	585	310
Frog neuron (potassium)	8.6	181	36
Frog neuron (sodium)	15.8	230	130
Gramicidin (sodium)	12.6	265	286
Sarcoplasmic reticulum (sodium)	20 (10 $\mathring{A}$ length)	2600	4900
Porin (sodium)	65	2250	1900

[a] Assuming channel length of 30 $\mathring{A}$ (or 10 $\mathring{A}$) and converting to 25°C. From Stein (1986).

just that given by Eq. (3.1), i.e., on the assumption that the diffusion coefficient of the ion in bulk water is appropriate for diffusion through the channel. Agreement between model and experiment is surprisingly good, and it seems that the simple scheme of a channel selecting largely on the basis of size accounts for at least some of the properties of these important ion channels. (We shall see in Box 3.4 that this agreement between our simplified model and the experimental data is to some extent fortuitous. Two factors, one increasing conductance and the other decreasing it, have been omitted in the calculations for Table 3.1 but are handled further in Box 3.4.) We still, however, have to account for the most interesting aspect of many of them, the detailed specificity according to which they discriminate between various ions. We proceed now to consider this problem.

Consider again the data for the end-plate channel, depicted in Fig. 3.5. The conductance values for the monovalent cations using this channel fall in the series $Tl^+ > NH_4^+ > Cs^+ > K^+ > Na^+ > Li^+$. This is just the series that describes the conductances for these ions in bulk water, and is almost inverse the sequence of their sizes in ionic crystals. These facts suggest that, within the end-plate channel, the ions behave as if they were still in bulk water! In water, all these ions are bound to the so-called water of hydration. These water molecules are bound to the ion more or less tightly depending on the radius of the ion (the smaller the ion, the larger is the field strength due the charge that it bears and the more tightly, and the more numerous are the water molecules bound). Thus, the smallest ions have the most water bound to them, forming loose structures that diffuse in water more slowly than those formed around the larger ions. The end-plate channel is thus wide enough to admit ions still bound to their water of hydration, and its selectivity between monovalent cations is determined by the size of their hydrated ions.

Box 3.4. Excluded Area and Partitioning: Two Further Factors That Influence the Behavior of Membrane Channels

The discussion about Eq. (3.1), which led to the calculations recorded in Table 3.1, neglected two important factors which act in different directions, one increasing, the other decreasing, conductances. Consider, first, the *excluded area effect*. Note the diagram in Fig. 3.6(a), where the cross section of a channel is depicted as a circle of radius R, while the diffusing ion is a smaller circle of radius r. The cross-sectional area of the channel is πR^2, while that of the ion is πr^2. The ion will pass through the channel only if it does not hit the channel wall. This will occur if the center of the ion enters the channel within the area given by the circle with radius $R - r$. Thus only a *fraction* of the total area of the channel is available for the ion. The rest of the area is excluded. The simple geometry of the model shows that the available

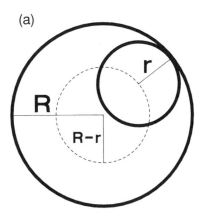

(a)

Fig. 3.6. Effects of excluded area and of partitioning on channel conductances. (a) Cross section of a channel, where R is the radius of the channel; r, that of the permeating species. (b) How the charge within a channel affects its conductance. The conductances of various derivatives of the end-plate channel (prepared by Numa and colleagues by site-directed mutagenesis; see text) are plotted on the ordinate against the charge calculated to be present within the channel. Replotted from the data of K. Imoto *et al.* (1988) *Nature (London)* **335,** 645–648.

(Continued)

area is given by the fraction $\pi(R - r)^2/\pi R^2$ or by $(1 - r/R)^2$. For the end-plate channel, for which we can estimate the cross-section diameter as 0.65 nm (see Fig. 3.7), and for the sodium ion which, when hydrated, behaves as if it has a cross-section diameter of about 0.45 nm, the available area is only 0.095, i.e., about one-tenth, of the total

(b)

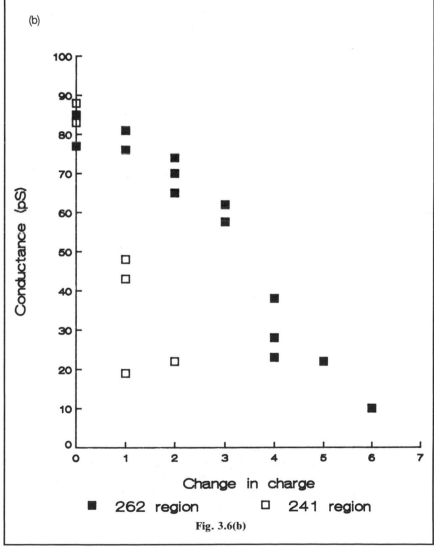

Fig. 3.6(b)

area used in the calculation for Table 3.1. The excluded area effect leads to a substantial decrease of channel conductance in this case (and in many other cases).

Acting in a contrary manner is the effect of ion *partitioning*. We have seen that the end-plate channel discriminates effectively between cations and anions, the latter having immeasurably small permeabilities through this channel. We postulated that this effect could be due to the presence of negative charges within the mouth of the channel. Support for this postulate comes from studies on the permeability of *uncharged* molecules through the end-plate channel. Urea and ethylene glycol are two nonelectrolytes that have much the same size as a hydrated sodium ion and can be shown to pass through the end-plate channel at a rate that is about one-tenth of that of sodium. Thus the partitioning of the cation within the channel acts so as to increase its permeability tenfold above that of a similar-sized uncharged permeant. Partitioning and excluded area affect permeability here in opposite but compensating fashion.

The postulated effect of partitioning has received further support from some fascinating experiments performed by Numa and colleagues. Using the techniques of molecular biology (see Box 3.3), they prepared polypeptide chains of the various subunits of the end-plate channel, substituted at defined positions in the polypeptide, at sites which carry negative charges. As the data depicted in Fig. 3.6(b) show, a reduction in the number of negative charges in the three rings of charges thought to be present in the end-plate channel, led to a graded reduction in conductance. In two rings that are thought to extend into the mouth of the channel ["262 region" in Fig. 3.6(b)], five negative charges had to be neutralized to bring about a tenfold drop in conductance. In the ring of charges that is thought to be at the center of the channel ["241 region" in Fig. 3.6(b)], a reduction of two negative charges was sufficient to lower the conductance tenfold.

Conductance data as a function of the radii of the permeating ions are available for two other channels in nerve membranes, one selective for potassium ions, the other for sodium ions. (We discuss the properties of these channels in more detail in Section 3.6.2.) Figure 3.7 shows the models that Bertil Hille developed for these three channels, giving in diagrammatic form the cross-sectional areas of the narrowest apertures that can account for the size-dependent selectivity of each of these three channels.

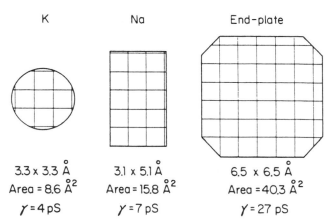

Fig. 3.7. Schemes for the effective size of various membrane channels. Hypothetical cross sections of three ion channels, drawn on the basis of ion permeability data. The grids are in units of 0.1 nm (1 Å); K depicts the potassium channel; Na, the sodium channel, End-plate, the end-plate channel. The single-channel conductance for each channel is stated below each figure, in picosiemens. Reproduced from T. M. Dwyer, D. J. Adams, and B. Hille (1980) *J. Gen. Physiol.* **75**, 469–492 by copyright permission of The Rockefeller Press.

The wide aperture of the acetylcholine receptor channel accounts for its low specificity. The potassium channel, narrowest of the three depicted, cannot allow a hydrated ion to pass through it. Any ion that is to pass through this channel must be stripped of its bound water of hydration. To do this, without replacing these bonds with others, would require an enormous input of free energy, which would result in a dramatic lowering of ion conductances. It seems that hydrophilic groups that line the walls of this channel (and other similar channels) can form transient bonds with the permeating ions, replacing the water that is bound to the ions in bulk water, and greatly reducing the energy requirements for permeation through the channel. It would appear that within the potassium channel, the ion-binding sites are somewhat weak. Thus they cannot replace the water that is so strongly bound to sodium and lithium. They can, however, replace the water that binds less strongly to potassium and to rubidium ions. Sodium and lithium, although small when naked, cannot pass through the potassium channel since they do not become stripped of their bound water.

The sodium channel, of intermediate width, selects for sodium over potassium by a mechanism that is still to be elucidated but which seems to involve a specific association between the channel walls and a single water of hydration that remains tightly bound to the sodium ion. A calcium channel present in frog heart muscle membranes possesses a rather special mechanism for ensuring its high specificity for calcium. The channel has two high-affinity, specific binding sites for calcium. When a calcium ion is bound to *one* of these two sites, no ion can get past it, as the channel is effectively blocked. Another calcium ion bound to the second binding site will, however, eject the first-bound calcium from its site, causing this ion to pass through the channel and out at the far side. In the absence of any calcium bound, i.e., at exceedingly low calcium concentrations, the channel allows monovalent ions to pass (even a cation as large as hydroxylammonium can get through the channel) and, in such circumstances, would not be described as a calcium-specific channel! It is only when calcium ions are present at near-physiologic levels, so that at least one calcium ion is (almost) always bound, that the high specificity for this ion becomes manifest.

3.4. AN EXPERIMENTAL INTERLUDE

3.4.1. Identification of Channels by Patch-Clamping

A typical cell membrane is furnished with numerous channels of different specificity, some passing sodium ions, others potassium, and still others calcium or even chloride ions. Many of these channels are regulated by signaling substances that pass from cell to cell, or by metabolic changes taking place within the cell, or by the ambient transmembrane potential. At different times in the life of the cell, often on an exceedingly short time scale, different channels may be open or closed, and the flow of ions entering or leaving the cell can alter from instant to instant. All these changes themselves affect the ambient membrane potential. Let us consider how such channels can be identified in cell membranes, and how the membrane potentials that they establish can be measured.

In the mid-1970s, Bert Sakmann and Erwin Neher developed a technique known as *patch-clamping,* which revolutionized our ability to identify individual channels in situ in cell membranes. Their technique is as follows (Fig. 3.8): Take a small glass pipette (a micropipette, about 1 μm in diameter at its tip) and, using a microscope and a micromanipulator, apply it to the surface of a cell. (To get a sense of the scale of the objects in question, recall that the cell itself will probably be 20–50 μm in diameter.) If small enough, the aperture of the pipette might cover one or only a few channels. By applying suction to the pipette, one can draw into the pipette a piece of the cell membrane and can detach this whole piece from the membrane itself. One now has a patch of membrane that can seal off

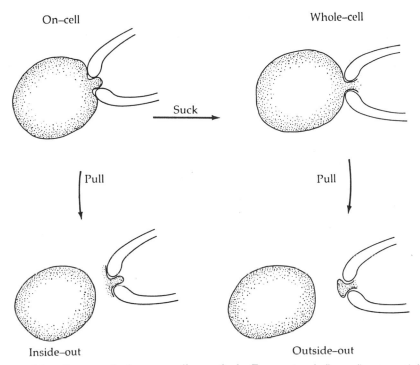

On–cell

Whole–cell

Suck

Pull

Pull

Inside–out

Outside–out

Fig. 3.8. Some patch-clamp recording methods. Four protocols for performing patch-clamp experiments. In each case, the cell is seen to the left, the micropipette to the right. In the top two figures, the pipette is pressed against the membrane to form a high resistance [in gigaohms (gΩ)] seal and left in place, enabling the conductance of a tiny patch of membrane (left) or the whole cell (right) to be measured. In the lower half, the pipette is, in addition, pulled off the membrane, isolating the small piece of membrane held across the pipette mouth. This will be inside–out or outside–out as depicted. In all cases the membrane held within the jaws of the pipette is the barrier for transmembrane movement of current so that suitable electrical circuitry can enable the conductance of this piece of membrane to be measured. Taken from Hille (1984), with kind permission.

the end of the micropipette. The pipette can be carried from one incubation bath to another, and immersed in solutions of different composition. Other solutions can be put into the body of the micropipette. In this way one can accurately control the composition of the fluid bathing *each* face of the patch of membrane. Finally, the electrical potential across the membrane patch can be clamped at any desired value, and the current flowing across the patch measured. All this, of course, requires very sensitive current-measuring devices and requires that the patch forms an extremely tight (high resistance, megohm or gigaohm) seal across the mouth of the micropipette. The glass must be highly polished and totally clean to enable the membrane to seal tightly to the glass. If the experimenter is skillful, the patch will contain one or a few channels and, if the concentrations of channel-regulating substances and the applied voltage are chosen correctly, the record of the current flowing will (Fig. 3.9), just as in the gramicidin experiment of Fig. 3.2, show a series of fluctuations, from the height of which one can calculate the *conductance* of the chan-

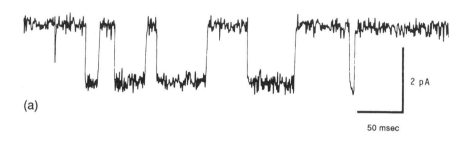

(a)

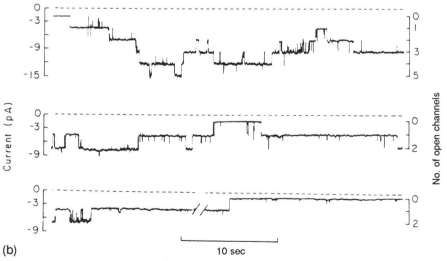

(b)

Fig. 3.9. Patch-clamp recordings. (a) Patch-clamp records of currents through single end-plate channels, from toad skeletal muscle, measured in a sodium saline solution at 11°C with a transmembrane voltage of −70 mV. Taken, with kind permission, from Barry and Gage (1984). (b) Currents through single calcium channels from rat heart muscle, measured in a sodium saline solution containing 6 μM Ca, 26°C, −70 mV. The records show up to five channels open simultaneously. Taken, with kind permission from D. Colquhoun *et al.* (1981) *Nature* **294,** 752–754.

TABLE 3.2

Conductances of Some Ionic Channels[a]

System	Specificity	Conductance (pS)
Gramicidin channel	Cations	4–7 (V_{max})
Ligand-gated channels		
Acetylcholine receptor	Cations	4–50
Glutamate receptor	Cations	120
Glycine receptor	Anions	30–70
GABA receptor	Anions	10–20
Sodium (voltage-gated) (nerve, muscle)	Na	4–20
Potassium (voltage-gated) (nerve, muscle)	K	2–20
Potassium (Ca-activated)		
Nerve cells	K	10–20
Red blood cells	K	20
Sarcoplasmic reticulum	K, others	180–240 (for K)
Chromaffin cells	K	180
Calcium (many cell types)		
T (Tiny)	Ca	8 (Ba)
L (Large)	Ca	25 (Ba)
N (Neurons)	Ca	15 (Ba)
Chloride channel (muscle)	Anions	400

[a] From Hille (1984); Stein (1986); Tsien (1987).

nel. Many channels have now been studied in this way using a wide variety of animal and plant tissues, and Table 3.2 lists some of these values. Channels opened (or closed) by chemical transmitters, by calcium, by voltage, and by protein phosphorylation have all been identified, their conductances measured, and their sensitivity to various activating and inhibiting substances explored.

One can also identify channels (less dramatically perhaps) by isolating the channel-forming molecules from cell membranes and reconstituting the channels back into lipid bilayer systems (such as were used in Fig. 3.2 for gramicidin) or into phospholipid vesicles ("liposomes"). In a combination of this technique with patch-clamping, one can seal the reconstituted artificial bilayer containing the putative channel across the mouth of the micropipette and study the behavior of the channel as before. Finally, as we saw in Box 3.3, one can now introduce the messenger RNA coding for channel subunits into the toad oocyte expression system. This leads to the expression of the newly synthesized channels in the membrane of the oocyte, from where they can be removed as a patch and studied as single channels.

3.4.2. Measurements of Membrane Potential by Using Intracellular Microelectrodes or by Following Dye Distribution

The ability to measure directly the electrical potential that develops across a cell membrane in a living cell was one of the great achievements of modern biophysics. The development of this technique (in the hands of the American biophysicist Kenneth Cole, and the Britishers, Alan Hodgkin and Andrew Huxley) required the invention and perfection of microelectrodes, small enough to be inserted into cells, or along the length of a nerve axon, without damaging the cell. The microelectrodes themselves are found in many different configurations; a frequently used one is depicted in Fig. 3.10.

A second handy method for measuring membrane potentials makes use of the fact that a charged molecule will tend to distribute itself across a membrane according to the prevailing transmembrane potential. If the molecule in question has a sufficiently high permeability, it will reach an equilibrium distribution across the membrane, given by Eq. (2.5). Suitably charged molecules need to be hydrophobic if they are to cross the cell membrane. Substances that have been used include the thiocyanate ion (SCN^-) and tetramethylphenylphosphonium ($TMPP^+$), the former for measuring potentials that are positive inside, the latter, for those negative inside. Thus if we can measure the ratio of concentrations on the two sides of the membrane, we can estimate the membrane potential $\Delta\psi$. It is convenient to use a fluorescent dye molecule as the probe of the membrane potential, since fluorescence measurements can be performed rapidly and with very high sensitivity. Only small amounts of the dye need to be added to the system, insufficient to perturb the membrane potential itself.

In many cases, the signal given out by a solution of such a fluorescent dye depends strongly on the dye concentration. At high concentrations, dye molecules interact with one another, quenching the fluorescence and reducing the signal. Thus, if the dye accumulates within the cell as a result of the prevailing transmembrane potential, its fluorescent signal is reduced (Fig. 3.11). For the method to work, it is essential that the dye be able to cross the membrane in its charged form. We shall see in Section 5.2.5 that fluorescent dyes can also be used to measure transmembrane pH gradients. This method is very convenient for measuring rapid changes in membrane potential (Fig. 3.11), but it has to be calibrated against known potentials before the light signal can be translated into absolute readings of membrane potential.

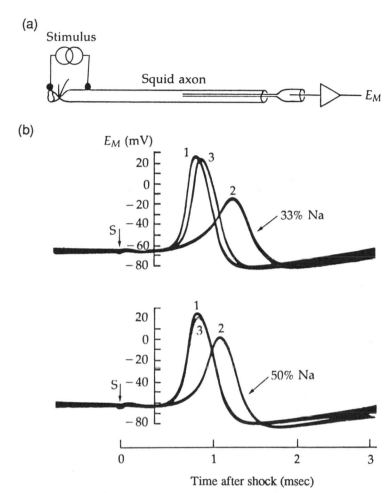

Fig. 3.10. Intracellular recording of transmembrane potential changes. (a) A microelectrode inserted along the length of a squid axon is shown schematically. The axon is stimulated by the device on the left and the signal is recorded by the inserted electrode. Part (b) shows the actual recording made. The potential is recorded on the ordinate as a function of time after the stimulatory shock. Recordings 1 and 3 show the signal with normal seawater bathing the axon. In record 2, the concentration of sodium is reduced to one-third. The experiment, indeed, was the first to show that the potential propagated along a nerve (the action potential, see Section 3.6.2) depends on sodium. Taken, with kind permission, from A. L. Hodgkin and B. Katz, *J. Physiol.* (*London*) (1949) **108,** 37–77.

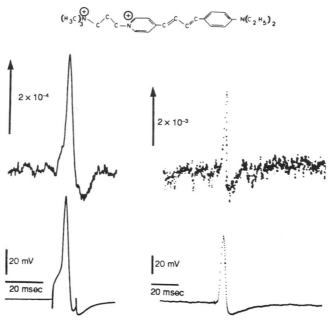

Fig. 3.11.   Optical recordings of transmembrane potential changes. The potential-sensitive red fluorescent dye RH-461 (structure in top figure) was used to measure the action potential (see Fig. 3.10 and Section 3.6.2) in a leech central neuron. The middle two records show fluorescent signals measured in this nerve after its stimulation; the record on the left is a computed average of five records, one of which is shown on the right. In comparison, the bottom traces are made by direct intracellular recordings as in Fig. 3.10. Taken, with kind permission, from A. Grinvald *et al.* (1987) *Biophys. J.* **51,** 643–651.

For many animal cells (see Table 2.2), the membrane potential is 50–90 mV, inside negative. For red cells, values of −10 mV are often found. In sections 3.5 and 3.7 we discuss the origin of this potential in detail, taking as an example the potential developed across the membrane of the squid axon.

3.5. DIFFUSION POTENTIALS: GOLDMAN–HODGKIN–KATZ EQUATION

In Section 2.2, when we discussed Eq. (2.5) we saw that, in a situation where there is only one ion diffusing across a membrane, a transmembrane potential will be set up that, when equilibrium is reached, is a function of the ratio of the concentrations of the diffusing ion at the two sides of the membrane. What happens if more than one type of ion is able to diffuse, that is, if several of the channels that we have listed above are simultaneously open in a particular cell membrane? If the ion concentration ratios are not exactly those given by Eq. (2.5),

$$\Delta\psi = RT/ZF \ln (S_{II}/S_I) \tag{2.5}$$

what potential develops? Some average potential? This is an important question, for much of the signaling between nerve cells (for instance, that which goes on in our brains) results in the opening and closing of ion channels and the summation of the resulting membrane potentials.

The problem is solved for us by the Goldman–Hodgkin–Katz (GHK) equation, which relates the overall membrane potential to a *diffusion potential* contributed by each ion, which is determined by the concentration of the diffusing ion at each membrane face and by the permeability of the ion across the membrane in question. Essentially, each ion tends to develop a membrane potential given by its concentration ratio across the membrane. The extent to which it reaches this potential depends on what fraction of the total amount of charge carried across the membrane is carried in the form of that particular ion, i.e., on its permeability and its concentration.

It would be too difficult for us to *prove* the GHK equation here [see the book by Nobel (1974) or some other advanced text on electrophysiology]. Note merely that the proof requires one to calculate for each ion in the system the current carried by that ion, as a function of the applied voltage. One then sets the *sum* of all the currents equal to zero to determine the transmembrane voltage at which no current flows. In deriving the GHK equation one has to assume that (1) the ions flow independently of one another, (2) the permeability coefficients for any ion are constant, and (3) the voltage drops linearly across the membrane. Assumptions 1 and 2 are invalid in all cases where an ion blocks the movement of other ions across the membrane, that is, when the ions compete with one another for movement within the channel and the channel becomes saturated. This situation is often found and it is, indeed, the study of deviations from the GHK equation that has led to many of the major advances in our understanding of electrophysiology. Assumption 3 implies that the electric field is constant across the membrane and, therefore, the GHK equation is often referred to as the "constant-field" equation.

Instead of a proof, we will simply state the GHK equation and show that it makes sense. The equation is commonly used in the following form, where the terms in p are the permeabilities of the ions in question and [K], [Na], and [Cl] are the concentrations of these three ions, at sides I and II of the membrane, as indicated:

$$\Delta\psi = RT/ZF \ln \left\{ \frac{p_K[K_{II}] + p_{Na}[Na_{II}] + p_{Cl}[Cl_I]}{p_K[K_I] + p_{Na}[Na_I] + p_{Cl}[Cl_{II}]} \right\} \qquad (3.2)$$

To understand Eq. (3.2), consider some special cases:

1. When only one ion can permeate (say, K), p exists for that ion but is zero for all others. Equation (3.2) reduces to Eq. (2.5) as it should.

2. When one ion (say, K) is *overwhelmingly* permeant, all values of p other than for that ion are insignificant and, once again, we recover Eq. (2.5)—which we should do, since the other ions play an insignificant part. [What this means is that if we have a cell membrane with many parallel channels for the different ions, the membrane potential across that membrane will reflect the concentration ratio of that ion whose permeability is overwhelming. The dominant ion (and hence the membrane potential) will vary as different channels with different ion specificities are opened and closed by the signals that the cell receives.]

3. It follows, also, that an ion that is present at a low concentration relative to other permeant ions will contribute very little to the potential, although it might well be as permeant as the other ions. The current flow (and it is this parameter that determines the potential) depends on both permeability and the absolute concentration of an ion.

We have written Eq. (3.2) in terms of those ions whose concentrations dominate most cells. Calcium and magnesium also contribute to the po-

---

Box 3.5. Using the GHK Equation to Determine the Ionic Specificity of Membrane Channels

Consider Eq. (3.2) for the situation where one ion, X, is present as the sole ion at one face of the membrane and another ion, Y, is the sole ion at the opposite face. The potential across the membrane will be zero when the concentration ratio of the two ions is the reciprocal of the ratio of permeability coefficients, i.e., where

$$p_X/p_Y = [Y_{II}]/[X_I] \qquad (B3.3.1)$$

Thus by determining the transmembrane concentration ratio that gives zero membrane potential, one can measure the permeability ratio for two ions. It was by using this technique that researchers obtained much of the important information on the ion specificities of various channels that we discussed in Section 3.3 (see also Fig. 3.5).

TABLE 3.3
Goldman–Hodgkin–Katz Equation Applied to Squid Nerve[a,b]

Ion	Concentration (mM) in		Relative permeability	Calculated equilibrium potential (mV)
	Cytoplasm	Seawater		
K	350	10	1	−91
Na	50	425	0.08	+55
Cl	41	560	0.19	−67

[a] Data from R. A. Sjodin, *in* Blaustein and Lieberman (1984). Cited in Chap. 2.

[b] $\Delta\psi$ calculated from Eq. (3.2) = −54 mV. $\Delta\psi$ measured experimentally = −60 mV.

tential in principle but, in practice, their concentrations are low and their permeabilities, except in certain special cases, generally far lower than those of the monovalent ions. Notice, too, that in Eq. (3.2) the chloride ion concentration ratio is reciprocal to those of potassium and sodium. This, too, is to be expected. Chloride ions diffusing down their concentration gradient will carry negative charge with them and charge the membrane negatively at the side *to which* they are diffusing.

Box. 3.6. Junction Potentials

The implications of the Goldman–Hodgkin–Katz equation apply to the flow of ions between any two phases. Thus at the junction between a microelectrode and the cytoplasmic solution into which it is placed, a diffusion potential will arise from the difference in ion compositions of the fluid within the electrode and that in the cytoplasm. This is the *junction potential,* which, if not adequately corrected for, can lead to serious errors in the measurements of membrane potentials. Note, however, that in aqueous media the diffusion coefficients of potassium and of chloride ions are almost identical. Thus if they are held at the same high concentration within the electrode, each will carry the same absolute current out of the electrode but each will carry a current of opposite charge. Thus there will be no *net* flow of current, and no interfering junction potential will develop across the interface between electrode and bulk medium. Junction potentials are eliminated if a high concentration of potassium chloride is used as a bridge between the microelectrode and the cell or bulk solution.

Table 3.3 lists the concentrations of the major intracellular ions, potassium, sodium, and chloride for a squid nerve and also the ionic composition of the seawater used experimentally as a model for the extracellular fluid that bathes the nerve in the living squid. Table 3.3 also lists the experimentally determined permeabilities of the nerve cell membrane for those ions *relative* to the permeability for potassium ions. The student can check that substituting these values into the Goldman equation [Eq. (3.2)] leads to a value for the resting membrane potential of 54 mV, inside negative. The measured value is closer to −60 mV. The difference is the contribution of the sodium pump (see Section 6.1). Table 3.3 also records the *calculated* Nernst potentials for each ion, i.e., the membrane potential at which the ion would be in equilibrium across the membrane for the prevailing transmembrane concentration ratio (see Section 2.2). Note that potassium and chloride are reasonably near Nernst equilibrium, sodium far from being so. Section 3.7 shows how these ion ratios originate.

3.6. REGULATION AND MODULATION OF CHANNEL OPENING

We have seen that various factors control the activity of the ion channels of cell membranes, ensuring that the cell is sensitive to the wealth of signals (chemical transmitters, electrical events and, as we shall see, hormones) that the cell receives. Modern physiology enables us to reach a detailed understanding of the signaling process in a number of cases. We shall discuss a few examples.

3.6.1. The Potassium Channel of Sarcoplasmic Reticulum

Contraction and relaxation of muscle are controlled by the calcium ion content of the cytoplasm of the muscle cell. To bring about relaxation, calcium is taken up into intracellular vesicles (the sarcoplasmic reticulum). Liberation of calcium into the cytoplasm brings about a contraction. The flow of calcium out of the sarcoplasmic reticulum, on specific calcium channels, is extremely rapid. If these were the only channels in the sarcoplasmic reticulum membrane, the large membrane potential that such an ion flow would induce would establish an equilibrium [defined by Eq. (2.5)] with more calcium ions remaining within the vesicle than outside it. The calcium can be released from the vesicles only if this membrane potential is dissipated. It appears that the sarcoplasmic reticulum membrane contains, in addition, a high concentration of potassium channels that allows a massive flow of potassium across the membrane of the vesicle, indeed overwhelming the calcium-induced potential. These potassium channels offer a high-capacity route for ion flow parallel to the calcium channels and "short-circuit" them, just as a conducting wire will short-circuit the two poles of a battery.

Chris Miller and colleagues studied this potassium channel intensively. The channel protein can be isolated from the vesicles and reconstituted into artificial liposomes. These liposomes can then be induced to fuse with planar bilayer membranes, so that the potassium channel proteins can be introduced in controlled amounts into a bilayer. When in the bilayer, their activity can be conveniently studied as a function of the composition of the media bathing the membrane, and as a function of the voltage applied across the bilayer membrane. This potassium channel has a very large conductance (see Table 3.2), 10 times greater than the end-plate channel (discussed in Section 3.2, above), 100 times that of the potassium channel of the nerve. Its conductance is very precisely regulated in response to the needs of the cell. Figure 3.12(a) shows a plot of the channel conductance versus the potential applied across a bilayer containing many potassium channels. The data show that conductance is a smooth function of applied voltage at all three pH values studied. Further study gives much insight into how voltage can control conductances. Consider what happens when few channels are present. Just as in Fig. 3.2, the curve of current flow against time then shows large fluctuations that can be analyzed. Figure 3.12(b) is a histogram plot of the *number* of occasions that a channel was found to have a conductance of the value plotted on the x-axis. At a negative voltage, most channels are closed—a few records show single channels open. As the voltage across the membrane becomes more positive, the chances of one, two, and even three channels being simultaneously open are increased, but the conductance of individual channels is not affected. Thus, changes in voltage open and close channels, but do not alter the characteristics of the open channels. It appears that the channel can exist in only two conformations (open or closed), which differ in the disposition of an electric charge. A single negative charge is carried across the membrane *during the transformation* from the open to the closed conformation. [Figure 3.12(a) shows that the conductance state of the channel is also affected by the pH, so that altering the pH of the muscle cytoplasm will affect the ability of the sarcoplasmic reticulum to release calcium to the cytoplasm. This is only one of many examples showing that the intracellular pH is another, important regulator of channel activities.]

3.6.2. Sodium and Potassium Channels of Excitable Tissue

Nerve and muscle cell membranes contain numerous voltage-sensitive channels that are selective for sodium or for potassium ions. Their sequential opening and closing in response to the prevailing transmembrane

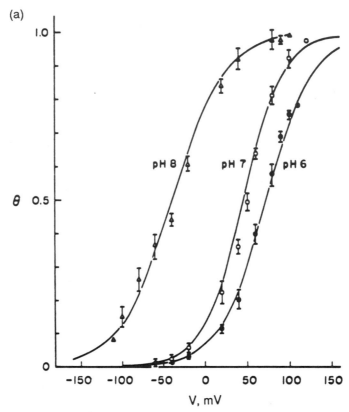

Fig. 3.12. Gating of a membrane channel: the potassium channel of the sarcoplasmic reticulum. (a) Shows how both voltage and pH control the opening of this channel. On the ordinate is plotted the fraction of channels that are open at the voltage specified on the abscissa. (b) Shows the histograms of the *number* of channels that are open at the different voltages noted on each subfigure, plotted against conductance, on the abscissa (see text). Reproduced from P. Labarca, R. Coronado, and C. Miller (1980) *J. Gen. Physiol.* **76,** 397–424 by copyright permission of The Rockefeller Press. (*Figure continues*)

potential is the basis for the propagation of the nerve signal along the axon and its spread across the muscle membrane. The sodium channel, in particular, has been well characterized, isolated, cloned, sequenced, and reconstituted into artificial phospholipid membranes. It is a single polypeptide chain of approximately 260,000-Da molecular mass. It is built of four repeating units, each rather similar to *one* of the subunit polypeptides of the end-plate channel that we discussed in Section 3.2. Each repeating unit is itself composed of six membrane-spanning stretches. One of these in each unit has the remarkable structure [see Fig. 3.13(b) and Box 3.7] of

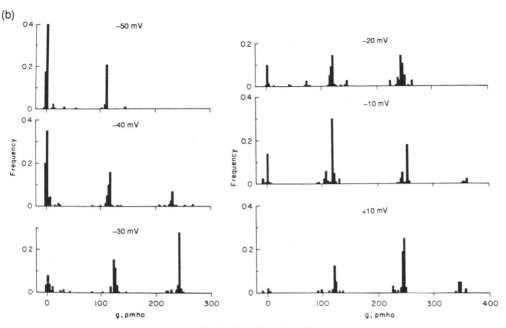

Fig. 3.12   (*Continued*)

a five- to sevenfold repeat of the triplet (–Arg–X–X–), where Arg (sometimes replaced by Lys) is the positively charged amino acid arginine (or lysine) and X is a hydrophobic residue. It is thought that this stretch of amino acids senses the voltage that gates (see Section 3.2) the sodium channel. Remarkably, the amino acid sequence of the sodium channel is similar to those of two other voltage-gated channels, a calcium channel and a potassium channel (see Box 3.7 and Fig. 3.13). It seems that, just as we saw in Section 3.2 for the chemically gated channels, there is a family of voltage-gated channels. There seems to be no common ancestry between the voltage-gated and the chemically gated channels.

Box 3.7. Molecular Biology of the Voltage-Gated Channels

The genes coding for voltage-gated sodium channels, calcium channels and potassium channels have been cloned and sequenced by a variety of interesting methods, giving results that are exciting and important.

(*Continued*)

First cloned, in the laboratory of Shosaku Numa, was the gene for the sodium channel from the electric eel *Electrophorus electricus*. A toxin from scorpions, the α-scorpion neurotoxin, binds with high affinity to nerve cell membranes, blocking the passage of the sodium current. This fact made possible the isolation of the sodium channel protein from detergent-solubilized membranes of the electric organ of the eel by identifying a protein in the membrane extract that bound the neurotoxin. Peptides were prepared from the sodium channel protein, by digestion with the enzyme trypsin. DNA sequences were then constructed that were possible coding sequences for certain of those peptides. As in the case of the end-plate receptor (see Box 3.3), cDNA clones were grown containing the sequences complementary to the messenger RNA from the electric organ cells. Antibodies were prepared (in rabbits) against the sodium channel protein, and the cDNA clones were screened to identify which produced proteins that reacted with the antibody to the known sodium channel. Four such clones were found and tested by hybridizing them against the DNA sequences constructed on the basis of the peptide digests. One clone proved positive; that is, it coded for a known portion of the sodium channel gene. But how were the researchers to locate the sequence that coded for the whole protein? DNA prepared from the identified clone was then used to search for sequences that overlapped with an end of *its* sequence, and those sequences were used, in turn, to search for sequences that overlapped with them. In this way, by extending out in both directions from the single original clone, DNA sequences composing the whole of the gene for the sodium channel were identified.

Figure 3.13(a) (top) depicts a possible arrangement of the polypeptide chain of the channel within the cell membrane, as deduced from hydropathy plots. (Refer to Section 1.4 for more on hydropathy plots.) The regions labeled I, II, III, and IV are each built up of six putative membrane-spanning sequences; the four regions are closely homologous with each other in their sequences. The sequence labeled 4 in each region (the S4 sequence) contains the five- to sevenfold repeating $-$Arg(or Lys)$-$X$-$X$-$ triplet that is thought to be the voltage-sensing element of the region. Figure 3.13(b) illustrates how such voltage-sensing might take place.

Next to be cloned, again by Numa and colleagues, was a calcium channel. Of the numerous types of calcium channels, one is blocked by the drug dihydropyridine and, as in the sodium channel just discussed,

(Continued)

the presence of such an effective channel-blocking substance led to the isolation of a protein that bound dihydropyridine, the preparation of peptides from the protein, and the eventual cloning of the gene coding for the protein. Its sequence [Fig. 3.13(a), middle] was found to be strikingly similar to that of the sodium channel, with the same pattern of four homologous regions, each containing six putative membrane-spanning stretches with many of the unique arginine-containing repeating sequences preserved in the S4 stretch. It has been shown that microinjection, into the cell nucleus, of a plasmid containing the DNA that codes for this calcium channel leads to the expression of calcium channel activity in the cell membrane. (In order to prove that this channel activity arose from the plasmid, the cells chosen as the expression system were muscle cells taken from rats that were suffering from muscular dysgenesis, in which such calcium channel activity is defective!)

Drugs that block potassium channels with high affinity have been harder to find, and this fact has hindered the isolation of potassium channels. Methods other than the use of known peptide sequences had to be developed to allow the gene for a potassium channel to be identified in DNA clones. Diane Papazian, Bruce Tempel, and colleagues used the technique known as "chromosome walking." They took advantage of the fact that the mutation *Shaker* in *Drosophila* results in a defect in a potassium channel. (*Shaker* mutants are so called since their legs shake when the mutant flies are anesthetized). The gene has been located by gene mapping at a particular point on the X chromosome (band 16F). A sequence of DNA that bound to the X chromosome in this region had been identified. Using this sequence, a "walk" was made in both directions along the chromosome by finding DNA sequences, cloned from the fly's genome, that overlapped with the ends of the sequence known to bind in the 16F region. The DNA clones so identified were then tested against DNA prepared from mutant flies in which the *Shaker* gene had suffered a deletion. In this way clones coding for the *Shaker* gene product could be identified as those that were not present in the mutant flies. Sequencing those clones gave the structure shown in Fig. 3.13(a) (bottom).

The sequence predicts a structure that is one-quarter of the size of the sodium and calcium channels, very similar to *one* of the four homologous regions in those channels. Expression in frog oocytes of the messenger RNA transcribed from the putative potassium channel

(*Continued*)

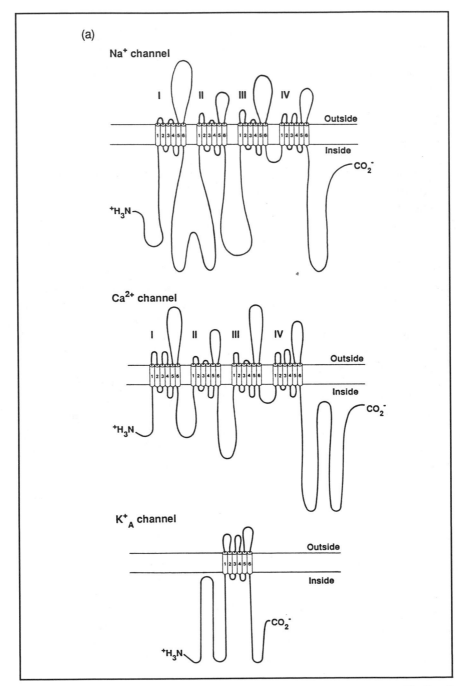

(*Continued*)

DNA gave rise to potassium channels in the oocyte membranes. A number of different potassium channels have now been cloned, all with the same general structure as that depicted in Fig. 3.13(a) (bottom). These different channel proteins may be responsible for the variety of potassium channels that the fly seems to possess.

The potassium channel is clearly the simplest of the three that we have been discussing. It appears that potassium channels arose first in evolution, being found in bacteria, unlike calcium and sodium chan- nels. These primitive channels were, presumably, composed of a single

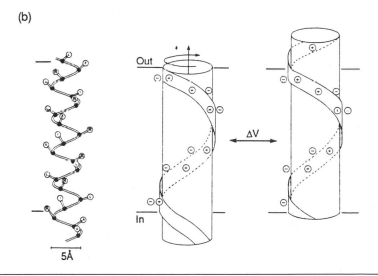

(b)

Out

ΔV

In

5Å

Fig. 3.13.   Structural models for the voltage-sensitive ion channels. (a) The three parts show possible models for the principal subunits of the sodium, calcium, and potas- sium channels, based on the hydropathy plots of their amino acid sequences (see Section 1.4). The first two channels are built of four repeating units, labeled I–IV, each com- posed of six putative membrane-spanning helices. The potassium channel seems to con- sist of only one of the units that repeat in the other two channel types. Part (b) shows a model for the voltage-sensitive conformation change that gates (opens and closes) the sodium channel. The six positively charged arginine residues in helix 4 of each large repeating unit of this channel are thought to be arranged in a spiral (the positive charges shown). Each positive charge is assumed to be held in place in the structure by a negative charge on the protein. When the transmembrane potential is altered, there is an increased force on these arginine charges and they are moved with respect to the remainder of the protein, the helix spiraling *through* the membrane. One turn of the helix moves a net single charge across the membrane. Taken, with kind permission, from Catterall (1988).

(Continued)

region of six membrane-spanning sequences. With the appearance of the protozoa, two rounds of gene duplication and the evolution of a new specificity led to the four-region polypeptides that we see as calcium channels. Evolution of the metazoans was associated with further evolution in the direction of a new specificity toward sodium and the appearance of sodium-specific channels still bearing the four-region motif.

How do the properties of the sodium channel and the potassium channel account for the propagation of the nervous impulse? Consider the model in Fig. 3.14, showing a nerve cell membrane that contains a voltage-sensitive sodium channel, a voltage-sensitive potassium channel, and another, unspecified but voltage-insensitive, pathway for potassium movement. In the uppermost picture, both voltage-sensitive channels are switched off. The membrane is in a "resting state" in which its permeability is low absolutely but relatively high for potassium ions. The membrane potential, as we saw in Table 3.3, is determined largely by that ion whose permeability dominates, in this case, by the potassium ions, and the measured resting membrane potential is, indeed, about −60 mV. If now the membrane potential is suddenly altered to inside positive [Fig. 3.14(b)] (how this happens will become clear in a moment), the voltage-sensitive sodium channel is switched on. The permeability of sodium now becomes dominant, and the membrane potential will be given by the ratio of *sodium* concentrations. These are low inside, high outside. This is how the potential reverses, becoming positive inside, as the sodium ions tend to rush in down their concentration gradient. Figure 3.10 is a record of such an "action potential," the changes in membrane potential that occur as the impulse passes a point on the nerve cell membrane. (In Box 3.8, below, we calculate how many sodium ions need to flow into the cell in order to reverse the membrane potential.) This change in potential has two effects. One effect is to switch on the next adjacent sodium channel [Fig. 3.14(c)]. In this way, a progression is achieved, a channel switching on an adjacent channel to propagate the signal down the axon. The other effect of the reversal of potential is to switch on the potassium channel. This is switched on with a delay and, responding to the ambient concentrations of potassium, restores the membrane potential to its original direction, inside negative [Fig. 3.14(d)]! The signal does not propagate in the reverse direction because any switching on of the sodium channel is overwhelmed by the potassium permeability that has been enhanced over that in the resting membrane. Therefore, the switching on of the potassium channel

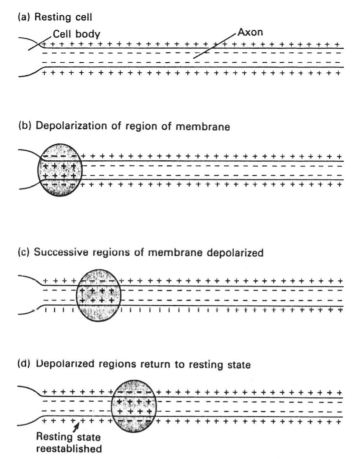

Fig. 3.14.   Propagation of the nerve impulse. (a) In the resting cell, the membrane potential is inside negative everywhere, since potassium channels are overwhelmingly the pathways for ion movement. (b) The membrane is locally depolarized, sodium channels open and this depolarizes the adjacent region of the membrane. (c) This depolarization itself depolarizes the next adjacent piece of membrane and the signal of depolarization is propagated onward. Potassium channels open with a delay, polarizing the membrane again, behind the region of depolarization. (d) The resting state is reestablished. Taken, with kind permission, from J. Darnell, H. Lodish, and D. Baltimore (1986) "Molecular Cell Biology." Scientific American Books, New York.

ensures that the axon has a refractory period after the impulse passes. Switching on and off of the nerve channels has a steeper voltage dependence than that depicted for the potassium channel of the sarcoplasmic reticulum in Fig. 3.12. The data are consistent with three or four charges driving the conformation change between the open and closed channels. (This is consistent with the presence of a voltage-sensitive amino acid sequence in *each* of the four large repeating units in the channel's primary structure, as discussed above and in Box 3.7.) These charges move with respect to the plane of the membrane, producing a tiny current known as the *gating current*. This current has been measured during the opening of a sodium channel in the nerve—a technical feat, since the gating current is 1000 times smaller than the current that flows through the channel itself.

3.6.3. The Cell-to-Cell Channel or Gap Junction

The cells in a tissue such as liver are a syncytium (i.e., the cytoplasms of the cells are in direct contact with one another), small solutes being able to pass freely from cell to cell through special channels, the gap junctions, that link cell with cell. These wide (2-nm-diameter) channels are regulated by calcium ions, being open at concentrations below $10^{-7} M$, while closed above $10^{-5} M$. The protein connexin, that constitutes these channels, has been isolated from cell membranes, purified, and studied by electron microscopy. Some details of the structure of the channel are known [Fig. 3.15(a)] and, in particular, the conformation change that opens and closes it [Fig. 3.15(b)]. The gap junction consists of six poly-peptide subunits, surrounding a central hole that forms the channel itself [Fig. 3.15(a)]. On the addition of calcium ions, each polypeptide's confor-mation about the axis of the hexamer alters. Each undergoes a slight twist [Fig. 3.15(b)], the overall effect of which is to narrow the central aperture, so as to render the whole molecule impenetrable. The effect of this struc-tural change is to ensure that the cell is sealed off from its neighbors if its internal calcium ion concentration rises. We know that the calcium ion concentration within cells is normally low, being at a level at which the gap junctions would be open. When, however, a cell is damaged, calcium enters the cell from the extracellular fluids, themselves rich in calcium. Finding themselves in a high calcium environment, the cell-to-cell chan-nels close, effectively sealing off the damaged cell from the rest of the organ with which it has been in communication.

3.6.4. Regulation and Modulation of Some Other Channels

Other channels are beginning to be studied with similar success. We know, for instance, that for the end-plate channel, two molecules of the neurotransmitter acetylcholine bind to the channel, and bring about its

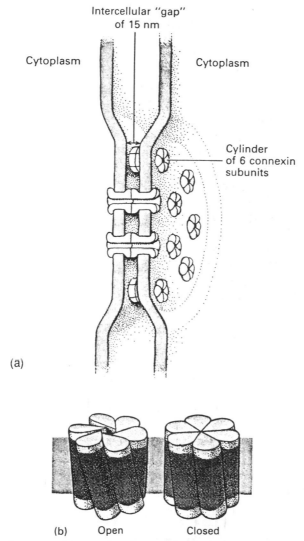

Fig. 3.15. Structure of the gap-junction channel. (a) Model of a gap junction, based on electron microscopic and X-ray diffraction analysis. Two plasma membranes are shown separated by a gap of about 15 nm. Both membranes contain cylinders of subunits of the protein connexin. The channel so formed across the gap connects the cytoplasm of the two cells. (b) A model for the regulation of opening of the gap-junction channel. Under the influence of calcium ions, the six subunits of the channel rotate, each about its own axis, so that the central gap between them is open (left) or closed (right). Taken, with kind permission, from (a) J. Darnell, H. Lodish, and D. Baltimore (1986). "Molecular Cell Biology." Scientific American Books, New York; (b) P. T. Unwin and G. Zampighi (1980) *Nature* **283**, 545–549.

transformation to an open conformation. When in this open configuration, the receptor switches very rapidly between a number of subconformations, not all of which have the same conductance. These subconformations are probably concerned in the fine regulation of the activity of the channel. The binding of inhibitory substances to the end-plate channel has been carefully studied, many of these being drugs of great pharmacologic importance (muscarine and atropine are two well-known examples). The acetylcholine receptor is subject to *desensitization,* the phenomenon in which the receptor becomes unresponsive to activation when it is exposed to acetylcholine for a long period of time (where "long" here means tens of milliseconds!). This *modulation* of the activity of the channel is clearly a phenomenon of some physiological importance. [Some scientists make a distinction between *regulation,* which best refers to the switching on and off or gating of channel activity (as brought about by some extracellular effector or the transmembrane voltage), and *modulation,* which refers to a change in the pattern of this activity as brought about by intracellular factors.]

Many cells contain enzymes known as *protein kinases,* which phosphorylate cellular proteins. This often leads to an alteration in protein activity. Certain of these kinases are themselves activated by cAMP or by Ca^{2+}, two substances that act as cellular "second messengers" in the mechanism by which the cell interprets the arrival of a hormonal signal into a change in the metabolic activity of the cell. (See Box 7.7, in Chapter 7, for an example of a "second messenger" affecting the activity of a chloride channel, a process impaired in this case in the hereditable disease, cystic fibrosis. See a cell biology textbook for further details concerning the mechanism of hormonal regulation). Ionic channels are, in many cases, targets for the modulatory action of these kinases. The end-plate receptor can be phosphorylated, and this phosphorylation is correlated with the desensitization of the receptor.

Calcium ions can directly modulate the activity of ion channels. We saw an example of this when we considered the gap-junction channel in the previous section, but there are numerous other examples. Many cell membranes contain a potassium channel that is opened when the calcium concentration of the cell rises from its normal level of less than 1 μM toward the level of 100 μM. This calcium-activated potassium channel may play a role in the regulation of cell volume, encouraging osmotically active potassium ions to flow out of the cell, thereby reducing the volume.

We might also mention a recently discovered class of channels that are modulated by the *tension* at the cell surface. Some of these channels are opened by the stretching of the cell surface; others are closed. These channels may also be involved in the control of the volume of the cell. (We consider volume regulation further in Section 7.1).

Finally, we must mention a family of membrane-associated proteins, the G proteins, which are vitally concerned in the modulation of cellular processes as a response to incoming signals. They can modulate ionic channels in important ways. One of these G proteins is *transducin*. This protein is found in the membranes of the light-sensitive neurons, the rod cells, and is involved in a cascade of molecular events that occurs when a photon of light is absorbed in the process of vision. The photon is absorbed by rhodopsin, which then activates transducin. Transducin regulates an enzyme that releases cyclic GMP bound to a sodium channel in the membrane of the light-sensitive cell, closing this channel and thereby modifying the transmembrane voltage (as we saw in Section 3.5, when we discussed the GHK equation). The change in transmembrane voltage is interpreted by the cell as a signal that light has been absorbed. This signal eventually reaches the brain to be recorded as a flash of light.

3.7. DONNAN POTENTIALS AND DONNAN DISTRIBUTION

Cells contain a large number of ionic species that are, to all intents and purposes, unable to cross cell membranes. These include proteins, nucleic acids, acids of intermediary metabolism, and nucleotides. In many cases, these ions are negatively charged, the "fixed anions." The presence of the accompanying small cations that complement the charge on these nondiffusible anions causes a diffusion potential (see Section 3.5) to be established across the cell membrane. Whenever a cation incipiently diffuses out of the cell, a separation of charges arises between this cation and the nondiffusible anions. The diffusion potential that is thus induced is such as to charge the cell membrane, interior negative. We can readily calculate the size of this potential. See Fig. 3.16.

Let the concentrations of diffusible cations and anions be symbolized by M^+ and A^-, respectively, that of the nondiffusible anion by N^-. Subscripts i and o refer to the concentrations inside and outside the cell. At each face of the membrane, the total concentration of negative charge must equal that of positive charge. We can write, therefore

$$M_o^+ = A_o^-; \qquad M_i^+ = N^- + A_i^- \tag{3.3}$$

We know, however, from Eq. (2.5), that at equilibrium the concentration ratio of the *diffusible* anion is the reciprocal of that of the diffusible cation, since the same membrane potential determines both of them. We have, therefore, in addition to Eq. (3.3), the following:

$$M_o^+ \times A_o^- = M_i^+ \times A_i^- \tag{3.4}$$

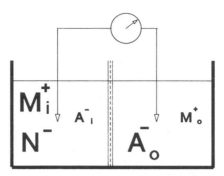

Fig. 3.16.   The Donnan distribution and Donnan potential. Two aqueous solutions i and o are separated by a membrane permeable to ions M^+ and A^- but impermeable to anion N^-, which is present on side i only. As a result, the concentration of the cation M^+ is increased at i over that at o, while the concentration of the anion A^- becomes low at i and high at o. The voltmeter measures the transmembrane potential that develops, which in this case is negative at side i and given by Eq. (3.5).

Solving these two simultaneous equations, and using Eq. (2.5), we find that

$$\Delta\psi = RT/ZF \ln(M_o^+/M_i^+)$$
$$= RT/ZF \ln(A_i^-/A_o^-)$$
$$= RT/ZF \ln(r) \qquad\qquad (3.5)$$
$$= RT/ZF \ln\{[1/(1 + N^-/A_i^-)]^{1/2}\}$$
$$= RT/ZF \ln[(1 - N^-/M_i^+)^{1/2}]$$

where r is the Donnan ratio. Equation (3.5) shows that a potential will be set up across the cell membrane, interior negative, due solely to the presence within the cell of nondiffusible anions. The size of this potential is dependent on the *ratio* of the concentrations of nondiffusible and diffusible anions (N^-/A_i^-), and it is diminished as the concentration of the diffusible anions becomes overwhelming. How many (or how few!) ions move from one phase to another when this potential is established is considered in Box 3.8.

The potential that develops across the membrane as a result of the presence of fixed anions within the cell is termed the *Donnan potential* after the physical chemist F. G. Donnan, working at University College, London, who demonstrated its existence (in membranes from nonliving systems). It is the major determinant of the transmembrane potentials that exist in all living cells (as recorded in Table 3.2) and that are so important as factors controlling ion distributions across cell membranes. The

Box 3.8. The Number of Ions Needed to Cross the Membrane in Order to Establish or Change the Membrane Potential

To calculate how many ions need to flow across the membrane in order to set up a particular transmembrane potential, we need first to understand the relation between the electric charge and potential. A substance such as a cell membrane, which is not a good conductor of electricity (as opposed to a copper wire), is known as a *capacitor*. It can act as an insulator, allowing a separation of electric charge to be established across it under the influence of an applied electrical potential. The charge and membrane potential are related by a quantity known as the *capacitance*. This is the measure of how much charge needs to be placed on the capacitor in order for it to develop unit potential. If the charge, Q, is measured in coulombs (96,500 coulombs are associated with one mole-equivalent of a monovalent ion), and the potential measured in volts (V), the capacity will be in farads (abbreviated F). Biological membranes have capacitances of size about $1 \mu F/cm^2$.

Take a typical animal cell of radius about 20 μm or 20×10^{-4} cm. This has an area of $4 \pi (20 \times 10^{-4})^2$ cm^2 and a capacity of that number of microfarads. To set up across it a membrane potential of the frequently found 60 mV will require a charge of $Q = CV = 4 \pi (20 \times 10^{-4})^2 \times 10^{-6} \times 60 \times 10^{-3}$ coulombs, which is associated with $4 \pi (20 \times 10^{-4})^2 \times 10^{-6} \times 60 \times 10^{-3}/\sim 10^5$ mol or 3×10^{-17} mol of ion, a very small number. The potassium content of such a cell, when the potassium is present at 140 mM in a volume of $4 \pi (20 \times 10^{-4})^3/3$ cm^3, is 4.7×10^{-12} moles. Thus to charge the membrane to 60 mV requires that as little as 6.4×10^{-4} percent of the cell's potassium be transferred across the cell membrane.

This ratio obviously depends on the *surface:volume* ratio of the cell. Let us calculate this ratio for a typical small axon of radius about 0.1 μm and containing 10 mM of sodium, where the inflow of sodium during the action potential changes the membrane potential from 60 mV inside negative to 60 mV inside positive. For a cylinder of radius r, the area to volume ratio is $2/r$. The ratio of the amount of sodium that must be moved in order to reverse the membrane potential, to the amount of sodium present within the cell, is $(120 \times 10^{-3} \times 2 \times 10^{-6})/(0.1 \times 10^{-4} \times 10 \times 10^{-1})$, or 2.4%. This is by no means an insignificant number and requires that the cell possess a very effective method of pumping out the sodium that has entered, if the axon is to maintain its ability to transmit an impulse. (We shall discuss this sodium pump in detail in Section 6.1).

presence of nondiffusible anions within the cell ensures that in the absence of other forces (and we shall see that these other forces can be dominant in certain important cases) cations will be concentrated within cells and anions will be excluded from them. This distribution of ions, brought about by the Donnan potential, is referred to as the *Donnan distribution,* symbolized by the parameter r in Eq. (3.5).

3.8. OSMOTIC CONSEQUENCES OF FIXED ANIONS WITHIN THE CELL

The presence of nondiffusible ions within the cell has another consequence that is exacerbated by the effects of the Donnan distribution. The total concentration of osmotically active material within the cell will always be greater than that in the extracellular solution, unless nondiffusible material is present there, too. The Donnan distribution adds to the problem because intracellular cation concentrations are increased above those in the extracellular solutions, adding to the osmotic pressure. The higher osmotic pressure within the cell will, as we saw in Section 2.7, cause a flow of water into the cell to equalize internal and external osmolarity. This dilution of the intracellular cation is followed by a renewed entry of cation in response to the Donnan potential. This, again, leads to further entry of water, osmotically. There is no point at which the cell interior and the extracellular solution are in equilibrium and the cell faces certain extinction by flooding, unless other forces supervene. What mechanisms have cells evolved to cope with this critical problem?

There seem to be two general solutions. One has been adopted by bacteria and plants. It is to accept flooding of the cell as a necessary evil, and to build a cell wall, a rigid skeleton around the cell, to hold the cell at the required volume. This means that the cell interior is under an enormous hydrostatic pressure offsetting the osmotic pressure of the nondiffusible cell constituents. Indeed, plant cells are held erect by turgor pressure, the hydrostatic effect of this osmotic inflow of water from the environment.

Animals have developed a different solution to the flooding problem, a solution that enables them to dispense with a rigid cell wall and to be actively mobile. An animal cell continually fights cell flooding by pumping out sodium ions as fast as they enter the cell. This ensures that sodium ions are effectively nondiffusible and kept *outside* the cell. This is a "double Donnan." The Donnan effect of intracellular nondiffusible anions is offset by holding sodium outside the cell, where its osmotic pressure relieves the threat of cell flooding. The animal cell must continually invest metabolic energy to pump out the sodium that continually enters down its concentration gradient. The use of energy is minimized by evolving a cell membrane that is as impermeable as possible to sodium—the less that enters, the less there is to pump out again—but a strictly impermeable membrane has never been achieved. As we shall see, animal cells spend a good proportion of their energy on pumping sodium ions. We discuss the mechanism of sodium pumping and its harnessing of metabolic energy in Chapter 6. Meanwhile, let us take a first look at a consequence of sodium pumping, the unequal ion distributions that result from it.

The ions that are *not* pumped out of the cell are subject to the Donnan potential brought about by the fixed anions, and distribute themselves according to Eq. (2.5) (see Table 3.2). Thus, potassium ions are generally at a higher concentration within than outside animal cells and are, indeed, the major intracellular cation. Concentration ratios of some 10-fold are the rule. Chloride ions are held outside the cell in reciprocal ratio to potassium. This is *not* an effect of pumping but a simple consequence of the Donnan potential that arises from the presence of the fixed anions within the cell. Magnesium ions tend to have a distribution in accordance with the prevailing Donnan relation (i.e., high inside the cell).

In contrast, both sodium ions and calcium ions, which are also being drawn into the cell by the Donnan potential, are continually pumped out of it and the cell maintains a low internal concentration of these two ions. The actual level of these ions that a cell develops at the steady state is a balance between the rate of outward pumping and the rate at which the ion leaks back inward. There is one ion pump specific for sodium (that also moves potassium inwards), another for calcium. Both are linked to the consumption of metabolic energy as we will see in Sections 6.1 and 6.3.

Potassium and magnesium ions are local signals of the cell interior; sodium and calcium ions, of the exterior. (We saw above how this was made use of by the cell-to-cell channel to ensure that cells are sealed off when they become leaky to calcium as a result of some injury). A rise in cell calcium is often a signal that some effector has interacted with the cell exterior, activating a specific channel and allowing the calcium ions to leak in down the prevailing concentration gradient. The subsequent rise in the intracellular concentration of an "extracellular" ion is, as we saw in Section 3.6.4, often the signal for a number of cell changes that alter the metabolic "set" of the cell to a new state.

We shall see in Section 5.3 that, in the case of sodium, the gradient of sodium ions built up by the sodium pump can be harnessed to pump into the cell metabolites that the cell requires for its day-to-day functions (or pump out of it those that it needs to get rid of). The overall inward leak of

Box 3.9. Time Course of Calcium Movements

It is instructive to calculate the time course of the changes in calcium ion concentration that a cell might undergo as a result of the opening of a calcium-specific channel and the subsequent action of the calcium pump. A conductance (see Section 3.1) of about 1 pS is typical for a calcium channel, bathed in an external medium containing calcium at 1 mM. Using the transformation factor reported in Section 3.1, we can calculate that 6.3×10^5 calcium ions will flow inward through such a channel (in the "open" state), each second. Taking Avogadro's number as 6×10^{23}, this flow is equivalent, roughly, to 10^{-18} mol of calcium. A red blood cell has a volume of ~10 pl (i.e., 10×10^{-12} liters). Therefore, each second, a single open channel will cause the calcium ion concentration of the cell to rise by $10^{-18}/10^{-11}$ mol per liter or by 0.1 μM. The resting level of free calcium ion in our red blood cell is about 50–100 nM, so *one* open channel would cause the intracellular calcium to rise 1- to 2-fold in one second, were all the calcium to remain free. (Much of it will, indeed, be bound to calcium-binding proteins in the cell, but, in many cases, such binding itself is binding to the proteins that signal the rise in internal cell calcium). What, now, of the outward pumping? Working at their maximum rates, the calcium pumps of the cell membrane (which we discuss further in Section 6.3) can extrude calcium from the cell interior at the rate of 30 ions per second. To expel the six million ions of calcium that enter through an open channel in ten seconds, a single pump would have to pump for 55 hr. In practice, the cell membrane contains many more calcium pumps than channels. A red blood cell, for example, contains 2000 calcium pumps in its membrane and yet takes 2 min to expel the calcium that a single channel admits each ten seconds.

sodium arises in part through unselective leaks and in part as a result of this harnessing of sodium flow to the coupled pumping of required metabolites. Finally, we saw above how the existing ion distribution across excitable cells is used to propagate signals along the cell membrane, using voltage-activated ion channels. In excitable cells, as in other animal cells, potassium ions are concentrated within the cell by the Donnan potential, while sodium is excluded from the interior by sodium pumping. An increase in the inward leak, brought about when a signaling effector opens a sodium channel, allows a rapid inflow of the ion down the maintained gradient. (See Box 3.8.) Restoring the ion gradients after the passage of the impulse is done by the sodium pump. The potassium gradient is maintained especially high in a nerve cell by increasing the concentration of nondiffusible anion. A nondiffusible anion, isethionic acid, is specifically synthesized by nerve cells, presumably to increase the concentration of fixed anion and hence increase the prevailing Donnan-induced potassium ion gradient. In contrast, in human (and many other) red blood cells, the internal concentration of potassium ions is kept above that due to the Donnan potential. This is due to the action of the sodium pump, which in these cells actually pumps potassium ions inwards as it pumps sodium ions outwards (see Section 6.1).

SUGGESTED READINGS

General

Hille, B. (1984). "Ionic Channels of Excitable Membranes." Sinauer, Sunderland, Massachusetts.

Stein, W. D., ed. (1984). "Ion Channels: Molecular and Physiological Aspects," Curr. Top. Membr. Transp., Vol. 21. Academic Press, Orlando, Florida.

Stein, W. D. (1986). "Transport and Diffusion across Cell Membranes," Chapter 3. Academic Press, Orlando, Florida.

Electrostatic (Born) Free Energy

Honig, B., Hubbell, W., and Flewelling, R. (1986). Electrostatic interactions in membranes and protein. *Annu. Rev. Biophys. Biophys. Chem.* **15**, 163–193.

Gramicidin Channel

Hladky, S. B., and Haydon, D. A. (1984). Ion movements in gramicidin channels. *Curr. Top. Membr. Transp.* **21**, 327–372.

Lee, W. K., and Jordan, P. C. (1984). Molecular dynamic simulation of cation motion in water-filled gramicidinlike pores. *Biophys. J.* **46**, 805–819.

Enzyme Kinetics

Segel, I. H. (1975). "Enzyme Kinetics." Wiley, New York.

Acetylcholine Receptor

Barry, P. H., and Gage, P. W. (1984). Ionic selectivity of channels at the end plate. *Curr. Top. Membr. Transp.* **21**, 2–51.

Guy, H. R., and Hucho, F. (1987). The ion channel of the nicotinic acetylcholine receptor. *Trends Neuro Sci.* **10**, 318–321.

Cloning and Molecular Biology

Imoto, K. *et al.* (1988). Rings of negatively charged amino acids determine the acetylcholine receptor channel conductance. *Nature (London)* **335**, 645–648.
Mishina, M. *et al.* (1984). Expression of functional acetylcholine receptor from cloned cDNAs. *Nature (London)* **307**, 604–608.

Acetylcholine Receptor Structure

Toyoshima, C., and Unwin, N. (1988). Ion channel of acetylcholine receptor reconstructed from images of postsynaptic membranes. *Nature (London)* **336**, 247–250.

Ionic Diffusion

Dani, J. A. (1989). Open channel structure and ion binding sites of the nicotinic acetylcholine receptor channel. *J. Neurosci.* **9**, 884–892.

Other Receptors

Schofield, P. R., *et al.* (1987). Sequence and functional expression of the GABA$_A$ receptor shows a ligand-gated receptor super family. *Nature (London)* **328**, 221–227.

Charge Effects on Channel Conductance

Begenisch, T. (1987). Molecular properties of ion permeation through sodium channels. *Annu. Rev. Biophys. Biophys. Chem.* **16**, 247–263.

Patch Clamping

Sakmann, B., and Neher, E. (1984). Patch clamp techniques for studying ionic channels in excitable membranes. *Annu. Rev. Physiol.* **46**, 455–472.

Fluorescent Dyes

Cohen, B., and Lesher, S. (1986). Optical monitoring of membrane potential. *In* "Optical Methods in Cell Physiology" (P. de Weer and B. M. Salzberg, eds.), pp. 71–100. Wiley (Interscience), New York.
Laris, P. C., and Hoffman, J. F. (1986). Optical determination of electrical properties of red blood cell and Ehrlich ascites tumor cell membranes with fluorescent dyes. *In* "Optical Methods in Cell Physiology" (P. de Weer and B. M. Salzberg, eds.), pp. 199–210. Wiley (Interscience), New York.

Goldman–Hodgkin–Katz Relation

Finkelstein, A., and Mauro, A. (1959). Physical principles and formalisms of electrical excitability. *In* "Handbook of Physiology" (J. M. Brookhart and V. B. Mountcastle, eds.), Sect. 1, Vol. I, pp. 161–213. Am. Physiol. Soc., Washington, D.C.
Goldman, D. E. (1943). Potential, impedance and rectification in membranes. *J. Gen. Physiol.* **27**, 37–60.

Nobel, P. S. (1974). "Introduction to Biophysical Plant Physiology." Freeman, San Francisco, California.

Potassium Channel of SR

Miller, C. *et al.* (1984). The potassium channel of sarcoplasmic reticulum. *Curr. Top. Membr. Transp.* **21,** 99–132.

Sodium and Potassium Channels of Excitable Tissue

Blatz, A. L., and Magleby, K. L. (1987). Calcium-activated potassium channels. *Trends Neuro Sci.* **10,** 463–467.
Butler, A. *et al.* (1989). A family of putative potassium channel genes in *Drosophila. Science* **243,** 943–947.
Catterall, W. A. (1988). Structure and function of voltage-sensitive ion channels. *Science* **242,** 50–61.

Molecular Biology of Voltage-Gated Channels

Noda, M. *et al.* (1984). Primary structure of *Electrophorus electricus* sodium channel deduced from cDNA sequence. *Nature (London)* **312,** 121–127.
Papazian, D. M., *et al.* (1987). Cloning of genomic and complementary DNA from *Shaker,* a putative potassium channel gene from *Drosophila. Science* **237,** 749–753.
Tanabe, T. *et al.* (1987). Primary structure of the receptor for calcium channel blockers from skeletal muscle. *Nature (London)* **328,** 313–318.

Cell-to-Cell Channel

Loewenstein, W. R. (1984). Channels in the junctions between cells. *Curr. Top. Membr. Transp.* **21,** 221–252.
Unwin, P. N. T., and Ennis, P. D. (1984). Two configurations of a channel-forming membrane protein. *Nature (London)* **307,** 609–613.

Calcium Channel

Tsien, R. W. *et al.* (1987). Calcium channels: mechanisms of selectivity, permeation and block. *Annu. Rev. Biophys. Biophys. Chem.* **16,** 265–290.

CHAPTER 4

Carrier-Mediated Transport: Facilitated Diffusion

We have seen that small molecules that dissolve well in organic solvents cross cell membranes easily and that small ions can often do so by using ion-specific channel proteins. Many of the nutrients that a cell requires for its existence are, however, hydrophilic and large, and cannot cross cell membranes by either of these two routes. Yet it is certain that the membranes of most cells contain very effective systems for taking up or extruding from the cell a great range of large, hydrophilic molecules (such as sugars, amino acids, nucleotides, and organic bases) that the cell needs or that it has to get rid of. In this section we shall discuss the specialized membrane transport systems that allow hydrophilic molecules to cross cell membranes. We shall show that these systems are, like the channels, proteins, but perhaps more akin to enzymes. They are embedded in the cell membrane and, by an enzyme-like conformation change, allow their specific substrates to cross the membrane from one face to the other. The membrane-embedded protein is called a *carrier*, *transporter*, or *uniporter*, and we say that the translocation of the substrate across the membrane occurs by *carrier-mediated (or facilitated) diffusion* or by *uniport*.

4.1. INHIBITION OF MEDIATED TRANSPORT SYSTEMS

The best evidence that some special component of the membrane is involved in a particular transport process comes from studies that show that transport can be *inhibited* by some reagent added to the system.

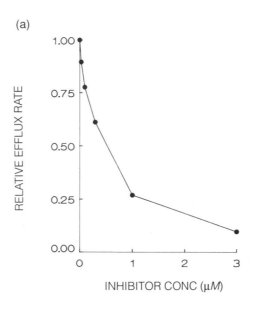

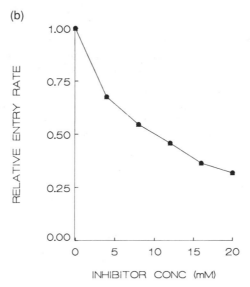

Fig. 4.1.   How inhibitors block movement of sugar molecules across red cell membranes. (a) Inhibition by cytochalasin B of the efflux of D-glucose (at 1 mM) from human red cells (20°C). Plotted from C. Y. Jung and A. L. Rampal (1977). *J. Biol. Chem.* **252,** 5456–5463. (b) Inhibition by maltose of the entry of D-glucose (at 0.5 mM, exchanging into cells preloaded with 200 mM D-glucose; see Section 4.2.4) into human red blood cells (at 20°C). Plotted from L. Lacko and B. Wittke (1982). *Biochem. Pharmacol.* **31,** 1925–1929.

Consider Fig. 4.1, which shows data on the transport of D-glucose across the human red blood cell membrane and how this is affected by the addition to the experimental system of (1) a drug, cytochalasin B, obtained from a fungus [Fig. 4.1(a)] or (2) a glucose analog, the disaccharide maltose [Fig. 4.1(b)].

There would be no reason to suspect that any molecule entering a cell by simple diffusion (Chapter 2) would be blocked by this drug or by a molecule such as maltose. Yet this is just what is found in the case of well-known enzymes. Many enzymes are inhibited by commonly used drugs, and many enzymes that act on glucose are inhibited by sugar analogs such as maltose. This sort of evidence led researchers to suspect that a membrane-bound protein might be involved in glucose transport. The postulated protein was described as *mediating* the transport of glucose and the protein termed the glucose "carrier" or "transporter."

The sugar galactose also competes with glucose, but in this case galactose *enters* the cell using the glucose transporter. Indeed, the glucose carrier can transport a wide range of substrates, all of them sugar molecules. The list includes hexoses, pentoses, and even a tetrose. On the other hand, disaccharides such as maltose seem to be too large to be transported. Aldoses, as well as ketoses, can use the glucose carrier, the former being in general transported more effectively. The glucose transporter does distinguish between optical isomers, preferring D-glucose over L-glucose. Thus the glucose carrier is *specific* for sugars but, apart from this preference for particular optical isomers, is not highly specific within the group of sugar molecules. (See Box 4.1 for an introduction to the molecular biology of the glucose transporters.)

Box 4.1. Molecular Biology of Glucose Transporters

The extent and detail of our knowledge of the glucose transporters in mammalian cells have increased dramatically in the last few years. This increase has come about as a result of the use of the newer techniques of molecular biology.

A glucose transporter was first isolated from human erythrocyte membranes by solubilizing these in detergent and then fractionating the resulting polypeptides, dissolved in the ionic detergent sodium dodecyl sulfate, on gels formed of polyacrylamide. The transporter was labeled by the attaching to it of a potent inhibitor, cytochalasin B [see Figs. 4.1(a) and 4.3(c)]. The protein was found to have a molecular mass of 55 kilodaltons (kDa) and to be a glycoprotein. The solubilized protein

---

(Continued)

could be incorporated into phospholipid liposomes (Section 1.3) with the recovery of its activity as a glucose transporter. This activity, like that of the transporter in its original location in the red cell membrane, was stereospecific (i.e., D-glucose was strongly preferred over the L-isomer).

Antibodies made in rabbits against the purified transporter were then used to clone the gene coding for the transporter. This was done by preparing *expression libraries* from a hepatoma (liver cancer) cell line and, in another study, from brain cells. In an expression library, cDNA (see Box 3.3) is prepared from a cell that expresses the protein to be studied. The different cDNA fragments in this preparation are then combined with a bacteriophage gene in such a way that the cDNA sequences of the cell will be expressed in the bacterial colonies that take up the cDNA fragments. In the expression library, the mRNA transcribed from the cDNA is expressed as protein sequences. These sequences are tested for their ability to react with antibodies prepared against the protein in question, in this case, the isolated glucose transporter. DNA from the clones so identified was sequenced and these sequences were put together to give the sequence of the whole transporter gene.

The resulting amino acid sequence of the transporter of the hepatoma is depicted in Fig. 4.2(a). Unlike most mammalian membrane proteins studied thus far, the glucose transporter does not carry a signal sequence (Section 1.5) that would command its insertion into the membrane during synthesis. Rather, the protein is synthesized on free ribosomes and only then inserted into the plasma membrane. The hydropathy plot for this amino acid sequence is depicted in the upper portion of Fig. 4.2(b). The molecule seems to consist of 12 membrane-spanning-helical sections, connected by hydrophilic sequences of which all but two are short. Treatment of the native, membrane-bound transporter with the proteolytic enzyme trypsin confirmed the correctness of the model of Fig. 4.2(a): the enzyme was shown to be able to digest substantial portions of those very parts of the transporter that Fig. 4.2(a) depicts as being present in the aqueous medium.

Although the DNA clones used in the sequencing of the glucose transporter had been prepared from a human hepatoma and from rat brain, the sequences so found were more than 97% identical with each other. That from the hepatoma seems identical with polypeptide sequences determined from the human red cell transporter. Thanks to the availability of the cloned DNA for the transporter, a study could be

(Continued)

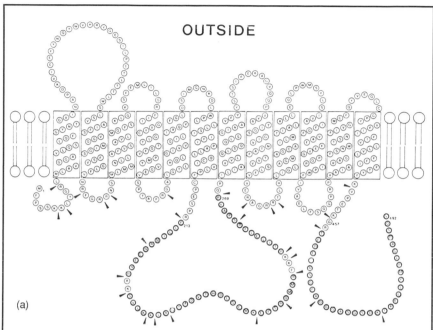

Fig. 4.2. Structure of the glucose transporter. (a) Model for the arrangement of the polypeptide chain of the glucose transporter of the human erythrocyte within the lipid bilayer, based on the hydropathy plot [see also Fig. 4.2(b)]. The letters within circles are the single-letter symbols for the amino acids. Arrows indicate the positions of cleavage by trypsin, acting from the cytoplasmic surface. Taken, with kind permission, from M. T. Cairns *et al.* (1987). *Biochim. Biophys. Acta* **905**, 295–310. (b) Hydropathy plots of the brain and liver glucose transporters. Taken, with kind permission, from Thorens *et al.* (1988). (*Figure continues*)

performed of the distribution of the relevant mRNA among the tissues of the body. The erythrocyte–hepatoma–brain glucose transporter was found to be expressed also in placenta, heart and, at lower levels, in intestine and kidney.

Strikingly, it is not expressed in normal (nonmalignant) liver nor in skeletal muscle or adipose cells, all cells that are actively engaged in glucose metabolism and known to possess glucose transporters in their membranes. A search was made for the "missing" transporter genes in these tissues. The method of search was to screen cDNA libraries of those tissues using the already identified DNA for the erythrocyte–hepatoma–brain glucose transporter in the hope of finding related cDNA molecules that would hybridize with sequences from this known

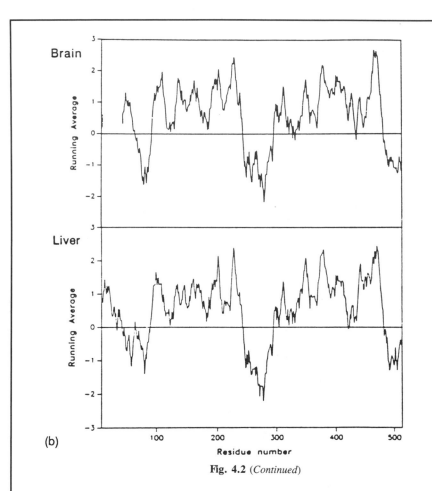

Fig. 4.2 (*Continued*)

DNA. Clones *were* found in normal liver and in adipose cells that hybridized with the DNA coding for the glucose transporter of the brain. The DNA so identified was eventually shown to come from two new genes. One of these new genes codes for a glucose transporter expressed in liver, and also in kidney and intestine. Its sequence is 55% identical with that of the transporter of the brain. Its hydrophobicity plot is given in the lower portion of Fig. 4.2(b). The other gene codes for a transporter expressed in adipose cells and also in skeletal and heart muscle, and its sequence is 65% identical with that of the brain's glucose transporter. There are thus at least three separate, but closely

(*Continued*)

related, species of molecules that take part in glucose transport in the various tissues of the body, and we must expect that each such transporter has its own special role to play in these different cells.

The role of one of these carriers is already becoming clear. The tissue distribution of the transporter of the adipose cell parallels the tissue distribution of sensitivity to insulin, the hormonal regulator of glucose transport in certain tissues of the body. Furthermore, an antibody made against a part of the transporter of the adipose cell manifested a membrane-bound protein in these insulin-sensitive tissues. It revealed this protein bound to plasma membranes in greater quantities when the tissues were stimulated by insulin! A beautiful study of adipose cells has been made by electron microscopy and by the labeling with gold particles of an antibody against the adipose cell transporter. This showed that the glucose transporter moved from a cytoplasmic location (attached to the Golgi apparatus) to the plasma membrane, following stimulation of the cells by insulin. The gene that codes for that glucose transporter which is expressed in adipose cells clearly codes for the insulin-regulatable member of the glucose transporter family.

Finally, this family is of a venerable lineage. It has been shown that certain sugar transporters present in bacterial cells share a good deal of sequence homology (~40%) with the mammalian glucose transporters. Interestingly, these bacterial sugar transporters are of the proton-activated type that we shall discuss in Chapter 5. The latter are known to bring about the accumulation of their sugar substrates against a concentration gradient, something that their mammalian sugar-transporting relatives cannot do.

4.2. KINETICS OF CARRIER TRANSPORT

Further insight into the mechanism of action of the postulated carriers came from experiments on the effect of the *concentration* of a substrate on its rate of transmembrane transport. A number of different experimental procedures can be used to study such effects of concentration. We discuss some of the most important procedures in turn.

4.2.1. The Zero-Trans Experiment

In the following simple experiment, the cells are initially free of the substrate. The initial rate of substrate entry is measured as the concentration of substrate is varied at the cis side. (Cis is the side from which the

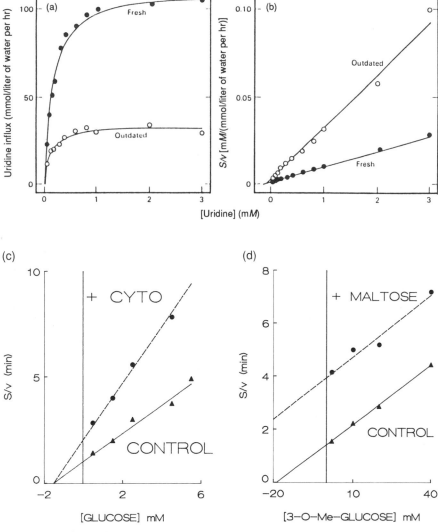

Fig. 4.3. Saturation and inhibition of carrier-mediated transport: (a) The v–S plot. Concentration dependence of zero-trans *uptake* of uridine into human erythrocytes (at 22°C). The solid circles depict data obtained using fresh red cells; the open circles, using stored blood. (b) A plot of S/v against S for these data [see text and Eq. (4.2)]. The data are best-fitted with K_m values of 0.15 and 0.12 mM for fresh and outdated blood, respectively, and V_{max} values of 111 and 34 mmol per liter of cell water per hour for these two cases. Figures 4.3(a) and (b) taken with kind permission, from S. M. Jarvis *et al.* (1983). *Biochem. J.* **210,** 457–461. (c) Inhibition of glucose *entry* into human red cells at 16°C by cytochalasin B. The lower line (triangles) shows data for experiments in the absence of the inhibitor; the upper line (circles), using 0.2 μM cytochalasin B. The plot is of S/v against S (see text) and

transport measurements are being made; trans, the opposite face of the membrane. Therefore, in this case, cis is the outside of the cell and trans is the inside.)

Figure 4.3(a) depicts a typical experiment using human red blood cells (fresh blood in the upper curve, outdated transfusion blood in the lower). On the y-axis is plotted the rate of transport of the nucleoside, uridine; on the x-axis the concentration of uridine in the medium in which the red cells are suspended. At low concentrations of uridine, the rate of uptake increases linearly with the uridine concentration. (This is to be expected on any mode of transport. The higher the concentration of substrate, the greater the number of impacts on each area of cell membrane, and the greater the number of substrate molecules that cross.) However, as the uridine concentration rises, the rate appears to rise less steeply, until at the highest uridine concentrations (~ 1 mM), the system becomes *saturated* and no further increase in the rate of nucleoside uptake can be obtained. This is *not* the behavior expected from simple diffusion. Nothing in the process of simple diffusion would lead one to predict saturation. It is, however, reminiscent of the behavior found for the movement of ions through the gramicidin channel (Section 3.1) and, as discussed there, is exactly the behavior also found for a large number of enzymes.

Were one an enzymologist, one would account for this behavior on the assumption that substrate combines with enzyme to form a *complex* in which the substrate is activated and undergoes the chemical change catalyzed by the enzyme. When *all* the enzyme is complexed with substrate, there can be no further increase in the rate of the reaction as more substrate is added. The system reaches its maximum velocity. In mediated diffusion, we can again assume that the substrate combines with the carrier to form a complex and, as a result, is enabled (in a way still to be elucidated in detail) to cross the cell membrane. As we saw in Box 3.2, the maximum velocity of transport (given by the horizontal portion of the curve of Fig. 4.3) is usually written as V_{max}, while the substrate concentration at which one-half this maximum velocity is obtained is defined as K_m, the Michaelis constant for the carrier–substrate system.

The mathematical equation that describes the form of the line in Fig. 4.3(a) in terms of v, the velocity of transport, $[S]$ the concentration of substrate, and V_{max} and K_m as defined in the preceding paragraph was

---

demonstrates noncompetitive inhibition. (d) Inhibition of the exchange of 3-O-methylglucose across red cell membranes at 16°C by maltose. The lower line (triangles) shows data for experiments in the absence of the inhibitor; the upper line (circles), using 40 mM maltose. The plot is of S/v against S (see text) and demonstrates competitive inhibition. Figures 4.3(c) and (d) are redrawn by the author from the original data of Basketer and Widdas (1978).

given in Box 3.2, and we reproduce it again here as

$$v = \frac{[S]V_{max}}{K_m + [S]} \tag{4.1}$$

which can be rewritten as

$$[S]/v = K_m/V_{max} + [S]/V_{max} \tag{4.2}$$

where V_{max} and K_m are the *kinetic parameters* that characterize the substrate–enzyme (or substrate–carrier) system.

The lines drawn through the points in Fig. 4.3(a) were calculated using Eq. (4.1) with a V_{max} of 111 mM hr^{-1} and a K_m of 0.15 mM for the upper curve, a V_{max} of 34 mM hr^{-1} and a K_m of 0.12 mM for the lower curve; these parameters were derived by applying a statistical best-fitting procedure to the data obtained experimentally. As we saw in Box 3.2, Eq. (4.1) is termed the *Michaelis–Menten equation*. Equation (4.2) is one of its *linearized* forms. From this straight line [see Fig. 4.3(b)], in which S/v is plotted on the y-axis and S on the x-axis, one obtains $1/V_{max}$ as the slope, while the intercept on the x-axis is equal to $-K_m$.

The maximum velocity (V_{max}) of an enzymatic or transport process will always be limited by the rate of some key chemical or physical process in the reaction being studied. The Michaelis constant K_m can be loosely interpreted in terms of the *affinity* that the enzyme shows towards the substrate. If K_m is low, the maximum velocity is reached at low levels of substrate consistent with a high affinity of enzyme and substrate. In both enzyme and transport studies, this simple interpretation of K_m as an affinity is complicated by the physicochemical equilibria that take place after the initial formation of the complex between enzyme–carrier and substrate. The identification of K_m as an "affinity" should in general not be made without an awareness of the possible complexities that can arise. We discuss this further in Section 4.6.

Some kinetic parameters for the sugar transport system of the red cell (that Fig. 4.1 depicted) are collected in Table 4.1. Very different values of K_m are found for the different sugars that the glucose carrier transports, while V_{max} values tend to be rather similar (see upper part of Table 4.1).

4.2.2. Competitive and Noncompetitive Inhibition of Transport

One can look more closely at the phenomenon of the inhibition of transport by also studying how the effectiveness of the inhibitor depends on the concentration of the substrate. We do this in Fig. 4.3(c) and 4.3(d)

TABLE 4.1

Kinetic Parameters of Sugar Transport in Human Red Blood Cells[a]

Different sugars[b]	Temperature (°C)	K_m (mM)	V_{max} (mM/min)
Glucose	37	4–10	600
	25	3	150
	20	1.6	45–200
Galactose	37	40–60	700
	20	12	150
Mannose	37	14	300–700
	20	7	120
Xylose	37	60	650
Ribose	37	2000	600
L-Arabinose	37	250	700
L-Sorbose	25	3100	125
Fructose	25	9300	125

Transport mode for D-glucose[c]	K_m (mM)	V_{max} (mM/min)	V_{max}/K_m (min^{-1})
Zero trans entry	0.5	0.38	0.74
Zero trans efflux	2.7	3.8	1.42
Equilibrium exchange (entry and efflux)	42	50	1.18
Infinite trans entry	0.8	23	
Infinite trans efflux	19	37	

[a] Outdated transfusion blood.

[b] Mostly zero trans entry experiments. Older data from Stein (1967) "The Movement of Molecules across Cell Membranes." Academic Press, New York.

[c] All at 0°C. From Wheeler and Whelan (1988).

for the inhibitors of the sugar transport system, cytochalasin B and maltose, that were used also in the experiments shown in Fig. 4.1. Three characteristic types of behavior are found. In the first, the inhibitor is found to affect only the maximum velocity of transport, leaving K_m [the intercept on the x-axis in a graph such as Fig. 4.3(c)] unchanged. This is *noncompetitive inhibition*, and is the behavior found for many drugs that act as inhibitors of transport. (In noncompetitive inhibition, the inhibitor can bind both with free carrier and with the carrier–substrate complex. Effectively, it removes functional carrier from the system.) Even at the highest concentration of substrate, inhibition cannot be reversed. In contrast, *competitive inhibition* can be reversed by higher concentrations of substrate. Since a higher concentration of substrate can give the same degree of transport activity as in the absence of inhibitor, the effective K_m is increased whereas V_{max} [the slope of the graph of, for instance, Fig. 4.3(d)] is not changed. (In competitive inhibition, the inhibitor can bind with free carrier but cannot bind with the carrier–substrate complex. This is because the substrate competes against the inhibitor for binding, just as inhibitor competes against substrate.) In a third type of inhibition, both V_{max} and K_m are affected. This is *uncompetitive inhibition*, a rare form in which the inhibitor binds only with the carrier–substrate complex.

The phenomena of inhibition and saturation that characterize carrier-mediated transport are often found, too, for diffusion through the specific ion channels, as was seen in Chapter 3. Indeed, it is certainly correct to consider diffusion through a channel system as a type of *mediated* transport, although not as one that uses a carrier system. There are excellent tests that *can* be used to distinguish between transport by a channel mechanism and by a carrier. These use some additional experimental procedures for measuring transport that will now be discussed.

4.2.3. The Equilibrium Exchange Experiment

The most reliable method of obtaining the kinetic parameters V_{max} and K_m of a transport system is to perform an *equilibrium exchange* experiment. Here, one suspends the cells in different concentrations of the unlabeled substrate, choosing a suitable range of substrate concentrations. After equilibrium is reached, a small quantity of an isotopically labeled sample of the same substrate, in an amount insufficient to disturb the equilibrium measurably, is added. At suitable time intervals, samples of the suspension are taken, the cells are separated from the medium (see Section 2.4), and the amount of radioactive label that has entered the cells is measured. The cells are in equilibrium before the addition of the label as well as during the uptake. Therefore, no change should occur in the state of the cells or their volume or the chemical environment. The *exchange* across the membrane of labeled for unlabeled sugar is the only thing that is happening, a fact that simplifies analysis enormously.

Whatever the mechanism of transport of the substrate, the uptake of label is a simple exponential function of time, if the cells are uniform in their behavior. A time course of the extent of filling or emptying of the cells with label (Fig. 4.4) must be a strict exponential curve and any deviation from this curve means that some unexplained artifact of measurement is present.

Because isotope is merely exchanging across the membrane, the time courses of influx and efflux of label should be mirror images of each other [Fig. 4.4(a)]. Again, a test of this prediction tests the validity of the experi-

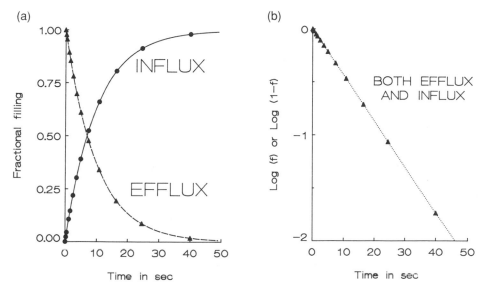

Fig. 4.4. Computed time course of equilibrium exchange efflux and influx. (a) Filling and emptying of the cell with labeled isotope in an equilibrium exchange experiment. (b) A semilogarithmic plot of the above curves—logarithm of fractional filling, f, against time.

mental system. Plots against time of the logarithm of f, the fractional filling (or emptying) of label [Fig. 4.4(b)], for different equilibrium concentrations, give the rate constants for efflux or influx, the quantities needed in Eq. (4.2) for obtaining V_{max} and K_m. Since the equilibrium exchange experiment does not in any way disturb the system, and since the method has such good internal checks, parameters found by this technique can be relied on to characterize the system in question. (Tables 4.1, 4.2, and 4.4 list kinetic parameters obtained for a number of different transport systems in the human red blood cell.)

4.2.4 Stimulation of Transport by Trans Concentrations of Substrate

Two further types of experiment also use radiolabelled substrates. Consider first what one might term a "low-cis" experiment. Using just the same experimental system as that used above, one loads cells with a labeled substrate to a low concentration inside the cell. One then transfers the cells to media containing various concentrations of unlabeled substrate (covering a range from well below to well above the value of K_m found in the equilibrium exchange experiment) and proceeds to measure

TABLE 4.2

Rates of Some Enzymatic Reactions and Some Transport Processes

System	Turnover number[b] (sec^{-1})	Temperature (°C)
Enzyme reaction[a]		
Carbonate dehydratase	10^6	
Acetylcholinesterase	10^4	
Urease	10^4	
Chymotrypsin	10^2–10^3	
Carboxypeptidase	10^2	
Transport process[c]		
Glucose (human red cells)	600	20
Galactose (human red cells)	900	20
Chloride (human red cells)	200	0
Phosphate (human red cells)	1.5	0
Uridine (human red cells)	2320	37
Lactose (*Escherichia coli*)	10	25
Na/K pump (kidney)	140	37

[a] From C. R. Cantor and P. R. Schimmel, "Biophysical Chemistry." Freeman, New York, 1986.

[b] The maximum velocity of an enzymatic or transport reaction divided by the number of enzyme or transporter molecules present. It is the maximum velocity per unit molecule of enzyme or transporter.

[c] From Stein (1986).

the efflux of labeled substrate. In the case of a carrier system, the initial rate of uptake of label as a function of the concentration of substrate outside the cells will be as depicted in Fig. 4.5(a).

As the substrate concentration outside the cell is raised, the rate of efflux of label is increased. When the extracellular substrate reaches very high concentrations, effectively infinite, efflux becomes maximal and labeled and unlabeled sugar exchange proceeds at a maximal rate. It must be emphasized that no channel would behave in this fashion toward uncharged substrate. If a channel is exposed to very high concentrations of substrate, i.e., when it has become saturated, substrate cannot leave from the side opposite to the high, blocking concentration. For all channels that have been studied, if trans substrate affects the flow, it reduces rather than increases the entry of labeled substrate. An experiment such as that depicted in Fig. 4.5(a) strongly suggests that the equilibrium between substrate and transport system at one face of the membrane is *shielded* from that at the other, so that what happens at one face of the membrane is independent of what happens at the other. Shielding the equilibrium at one face from the reactions occurring at the other requires some *reorientation* of the carrier during transport. Such evidence for reorientation is, indeed, the criterion for postulating the presence of a carrier system.

The presence of a carrier system is confirmed in the complementary experiment, the *infinite-trans* experiment. Here, the concentration of unlabeled substrate at the trans face of the membrane is set very high, well above K_m at that face and, effectively, infinite. The concentration of labeled substrate at the cis face is varied and uptake from this cis face is measured. Figure 4.5(b) shows results of a typical experiment. Again, one can define a maximum velocity and a K_m for this exchange and, again, no channel should give this unidirectional flux in the face of a saturating concentration of substrate at the trans face.

We shall see that the V_{max} and K_m terms of all the preceding experimental procedures are related to one another. To demonstrate this we shall have to analyze the kinetics of the carrier model.

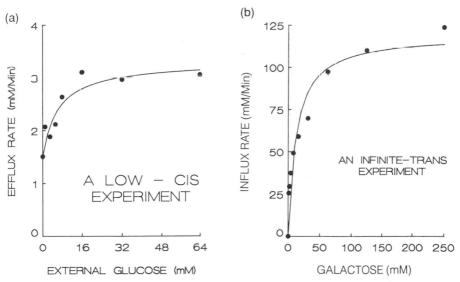

Fig. 4.5.   Stimulation of transport by a trans substrate. (a) A low-cis experiment: effect of external unlabeled glucose on the efflux of ^{14}C-labeled glucose from human red blood cells at 0°C. Initial internal glucose concentration was 2.6 mM. The ordinate depicts the sugar lost from the cells in 30 sec when the external concentration was as given on the abscissa. Redrawn from data of M. Levine *et al.* (1965). *Biochim. Biophys. Acta* **109**, 151–163. (b) An infinite trans experiment: uptake of ^{14}C-labeled galactose by human red cells at 25°C, loaded with 250 mM galactose. Ordinate is uptake per liter of cell water per minute, abscissa is concentration of the labeled sugar. Redrawn from data of H. Ginsburg and D. Ram (1975). *Biochim. Biophys. Acta* **382**, 369–376.

4.3. THE CARRIER MODEL

A formal diagram of the carrier model is given in Fig. 4.6. In this model, E symbolizes a carrier molecule that can exist in two conformations E_1 and E_2 which have their binding sites for the substrate S facing sides 1 and 2 of the membrane, respectively. Subscripts 1 and 2 for the substrate refer to its concentration on sides 1 and 2 of the membrane. The steps labeled g, with appropriate subscripts, refer to the changes in conformation of the carrier–substrate complex (the *loaded* carrier) that bring about the reorientation of its substrate-binding sites. Steps labeled k are the analogous reorientation rates of the *unloaded* carrier, while f and b, again with appropriate subscripts, refer to the rates of the *formation* and *breakdown* of the carrier–substrate complex, respectively.

The steps labeled g and k that involve the reorientation of the carrier (loaded and unloaded) would be expected to have rates typical for those of protein conformation changes, which range from a hundred to a million changes per second. (Compare Table 4.2, which lists rates of some protein conformation changes and of some carrier- and pump-mediated transport events. For more on pumps, see Chapters 5 and 6). The breakdown of the ES complexes would be expected to be even more rapid and, in the few cases where this has been directly measured, it does appear that the conformation changes of the carrier are rate-limiting for the overall reaction.

In many cases, the conformation changes of the loaded carrier are faster than are those of the unloaded carrier. Indeed, that is likely to be

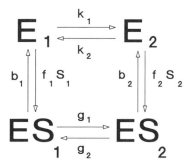

Fig. 4.6.   Schematic representation of the carrier or transporter model. The symbol E represents the transporter in its various forms; S, the transported substrate; ES, the transporter–substrate complex. Subscript 1 refers to side 1 of the membrane; subscript 2, to side 2. The variables f and b are the rate constants for the formation and breakdown of the ES complex, k for the transformations of the carrier that allow access of substrate to its binding sites at sides 1 or 2 of the membrane, while the terms in g describe the conformation changes of ES between forms ES_1 and ES_2.

the explanation of the data depicted as Fig. 4.5(a), where substrate at the trans side speeds up the rate of uptake of the label. To understand this, consider first what happens where there is no substrate at the trans face (i.e., in zero-trans conditions). If the *unloaded* carrier returns slowly, this will be the rate-limiting step of the overall scheme, and entry of label will be limited by the rate of return of free carrier. As the concentration of substrate is increased at the *trans* face, however, more and more frequently will the return journey be undertaken by carrier in loaded form. This rate may be, as we have suggested, faster. If so, the rate-limiting step will be accelerated. Uptake of sugar will become more rapid, until, when all the carrier at the trans face binds to substrate as soon as it dissociates, the overall rate of uptake will be limited by the rate of reorientation of the *loaded* carrier. The new maximum rate is now determined by this fast rate (faster than the rate of reorientation of unloaded carrier, but probably slower than the rates of breakdown of the carrier–substrate complex).

Thus a carrier system will often show trans acceleration of the uptake of label due to the presence of a substrate at the trans face. In contrast, a channel must (in the absence of electrical effects) always show trans inhibition, caused by blockage of the channel in *both* directions as the substrate concentration builds up at one or other face of the membrane.

4.4. VALINOMYCIN: AN ARTIFICIAL MEMBRANE CARRIER THAT WORKS BY A SOLUBILITY–DIFFUSION MECHANISM

We saw in Section 3.1 that the nonprotein ionophore gramicidin was a good model for the protein ion channels. In a similar fashion, the ionophore valinomycin (like gramicidin, an antibiotic isolated from a microorganism, but in this case built up of a trimeric repeating cycle of D- and L-valines, L-lactic acid, and D-α-hydroxyvaleric acid; see Box 4.2) is a good model for the kinetic behavior (although certainly not for the molecular mechanism) of the protein carriers. The nature of the ionophores, whether they behave as channels or as artificial carriers, can be distinguished by the effect on their properties of freezing the membrane in which they have been inserted. (As we saw in Section 1.6, lowering the temperature can bring about a phase transition in a phospholipid membrane, freezing it as a solid phase.)

The conductance of a gramicidin-doped membrane drops very little as the membrane freezes (although it might drop when the water within the channel freezes). This is consistent with the presence in such membranes of continuous channels within the gramicidin molecule. These channels, as we saw in Section 3.1, extend all the way across the membrane, form-

Box 4.2. Kinetic Analysis of Valinomycin as an Ion Carrier

Valinomycin has the chemical structure depicted in Fig. 4.7(a). It dissolves very readily in organic solvents and in cell membranes to which it imparts a high and selective permeability for potassium, rubidium, and cesium ions. Its mechanism of action can be accounted for by the carrier model of Fig. 4.6, on the assumption that the valinomycin spends most of its time in the membrane and, when unloaded, behaves like the carrier E in Fig. 4.6. In this form it combines with the ion, S in Fig. 4.6, to form the complex ES. Both E and ES diffuse across the membrane. Transport of S in the direction 1 to 2 involves the complete cycle of transport in which E at side 1 combines with S at this side, the complex ES_1 diffuses across to side 2. There S dissociates off to be counted as a molecule of S_2 and the free carrier diffuses back to side 1. Using a charge-pulse technique, Benz and Läuger were able to determine the four rate constants that describe the transformations of the carrier and substrate in this symmetric system.

In the charge-pulse technique, the membrane bilayer [Fig. 3.2(a)] containing valinomycin and situated between two identical solutions of a salt is charged to an initial voltage by a brief (10–100 nsec) pulse of current. At the end of the pulse the external charging circuit is switched off and replaced by a virtually infinite resistance. The membrane voltage decays by movement of charge across the membrane and by the redistribution of charges within it. From measurements of how the voltage changes with time [Fig. 4.7(b) shows time courses of the voltage decay, recorded at four different time scales], the values of all four rate constants of the symmetric version of the model of Fig. 4.6 can be found. One directly monitors the movement only of charged ES.

The curves in Fig. 4.7(b) show that the voltage decays in three phases. (The two top records are of the same phase but recorded on two time scales that differ by a factor of 5.) Each phase has its own amplitude. Roland Benz and Peter Läuger showed that measurements of the half-times of the three phases and of their amplitudes as a function of the concentration of the added valinomycin and of the transported cation could be analyzed so as to obtain separately the four rate constants of the symmetric model of Fig. 4.6, and also the *partition coefficient* for valinomycin in the membrane. The slowest process was the diffusion across the membrane of the unloaded carrier. This had a rate constant (for potassium at 25°C and the monoolein membranes that Benz and Läuger used) of 3.8×10^4 sec^{-1}. The rate constants for

(Continued)

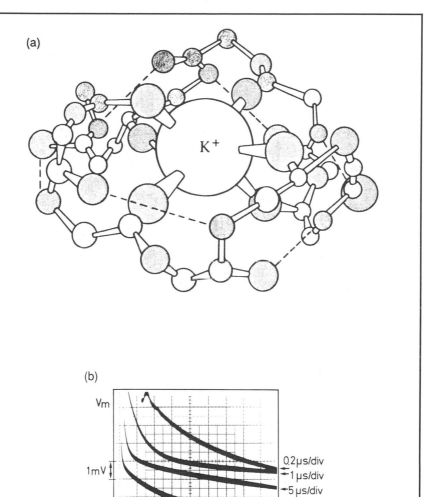

Fig. 4.7. Valinomycin: a diffusible carrier. (a) The structure of valinomycin. The central K^+ is surrounded by a cage formed by the amino acid side chains of the valinomycin. Taken, with kind permission, from F. M. Harold, "The Vital Force: A Study of Bioenergetics." Copyright © 1986 by W. H. Freeman and Company. Reprinted with permission. (b) The decay of voltage (V_m) after a charge pulse across a valinomycin-doped membrane made of monoolein/n-decane. Valinomycin at $2 \times 10^{-7} M$, in 1 M RbCl at 25°C. The voltage decay was recorded with different sweep times, as indicated on the right of the figure. Taken, with kind permission, from Benz and Läuger (1976).

(Continued)

transmembrane diffusion of the loaded carrier and for the breakdown of this carrier–potassium complex to free carrier and potassium were almost equal at $25 \times 10^4 \text{ sec}^{-1}$ and equal to the rate constant for formation of the complex (at 1 M potassium). The partition coefficient for the valinomycin–potassium complex between water and the membrane was 7.5×10^3, a high number. Clearly, the system is very well characterized by these valuable measurements.

Note that the maximum rate of action of the system is fixed by the rate of transmembrane diffusion of the unloaded carrier at $3.8 \times 10^4 \text{ sec}^{-1}$. This is about one-thousandth of the rate of the gramicidin channel discussed in Section 3.1. It is characteristic of carriers that they have a lower turnover number than channels. We discuss this point again in Section 4.9.

ing an effective route for transmembrane movement even across a frozen membrane. In contrast, the conductance of a membrane to which valinomycin is added is very steeply dependent on the state of the membrane. If the membrane freezes, the valinomycin cannot diffuse across it and the conductance can drop by four orders of magnitude in the temperature range over which freezing takes place. This experiment shows that, in order for valinomycin to work, it must *diffuse* across the membrane. It acts as a carrier because it complexes potassium ions, surrounding them in a cage of hydrophobic residues so that the whole complex can dissolve in the lipid membrane. It thus acts by a solubility–diffusion mechanism, quite different from the conformation change mechanism used by the proteins that act as carriers in cell membranes. Valinomycin is fairly specific for the carriage of potassium ions, preferring potassium over sodium by a factor of four orders of magnitude.

Peter Läuger has made very detailed kinetic analyses of the valinomycin-induced transport of potassium, fully confirming that it behaves according to the kinetic model of Fig. 4.6. Values for all four rate constants of the symmetric form of the model of Fig. 4.6 have been determined (see Box 4.2) using sophisticated kinetic analyses.

Valinomycin and other similar ionophores are tools of great importance for students of cell membranes who want to be able to control membrane permeability experimentally. The researcher can add valinomycin to a membrane in graded amounts and impose a desired level of potassium permeability. The diffusion potential of potassium can then be used to set a defined potential across the membrane (see Section 3.5).

4.5. TWO CONFORMATIONS OF THE CARRIER

In Section 4.2 we saw that the phenomenon of trans stimulation provides good evidence that two conformations of a carrier can exist, with
the interactions of substrate and carrier at the two faces of the membrane
being shielded from one another. There are a number of additional experiments that argue for two conformations. Return to the data depicted as
Fig. 4.3(c), where cytochalasin was used as an inhibitor of glucose *entry*
into human red blood cells. From Fig. 4.3, it is clear that glucose and
cytochalasin do not compete with each other, as the inhibition is of the
noncompetitive type. Yet when the effect of cytochalasin is studied on
glucose *efflux* from these cells, the drug is a competitive inhibitor! This
result was at first puzzling to researchers until Wilfred Widdas and
Richard Krupka with Rosa Devés, showed that it can be a direct consequence of the carrier model of Fig. 4.6, if we assume that cytochalasin
acts only at the inner face of the red cell. Figure 4.8 depicts the model that
these researchers proposed. When glucose is present at the inner face of
the cell membrane [Fig. 4.8(a)], it can compete with cytochalasin and the

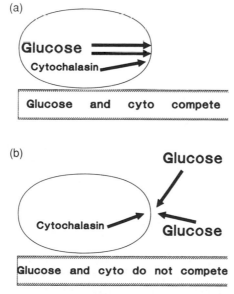

Fig. 4.8. Scheme for an experiment in which cytochalasin competes with sugar at only
one face of the membrane. (a) Glucose is present within the cell, so that glucose and
cytochalasin both have access to the binding site on the transporter and hence compete with
each other. In (b) the sugar is present outside the cell, so that cytochalasin and sugar do not
interact with the same binding site and therefore behave as noncompetitive inhibitors.

inhibition is competitive. When, however, glucose is outside the cell [Fig. 4.8(b)], it cannot compete with the binding of cytochalasin to the carrier, since this binding occurs at the inner face. The binding equilibrium is shielded from external glucose. Whether cytochalasin acts as a competitive or a noncompetitive inhibitor depends on the position of the substrate and this dependence on position requires that there is a reorientation of the transporter during the transport cycle.

Another type of experiment gives even more convincing proof of carrier reorientation. Sodium tetrathionate reacts with the glucose transporter, irreversibly inhibiting it. We can increase the rate of inhibition by adding maltose externally. Now maltose is a competitive inhibitor of glucose transport, but cannot cross the red blood cell membrane—whether or not it uses the glucose carrier. It thus binds to the carrier but prevents its reorientation, holding the carrier in its outward-facing form. That maltose increases the reaction rate with tetrathionate suggests that tetrathionate reacts with the carrier in this outward-facing form, and that maltose acts by *recruiting* carriers to their outward-facing conformation. In contrast, cytochalasin reduces the rate at which tetrathionate acts. This is also consistent with tetrathionate acting with the outward-facing conformation, since as we have just seen, cytochalasin interacts with the inward-facing conformation of the glucose carrier and, if present, will recruit carriers into the inward-facing conformation.

4.6. A DEEPER ANALYSIS OF THE KINETICS OF CARRIER TRANSPORT

The model in Fig. 4.6 lends itself readily to a formal kinetic analysis. Such an analysis can give a deep understanding of how the transport behavior of a carrier is determined by the kinetic parameters of the system, and these, in turn, by its rate constants.

4.6.1. Some Relations between the Transport Parameters for the Different Experimental Procedures

We saw in Figs. 4.3 and 4.5 that data from the zero-trans and infinite-trans experiments can all be plotted in the form of the Michaelis–Menten equation [Eq. (4.1)]. This is the case also for experiments using the equilibrium exchange procedure. For each experiment, then, one can define a V_{max} and K_m. Each experiment can also be performed in two directions, in-to-out or out-to-in. This gives us, in principle, six pairs of parameters that will characterize the carrier model of Fig. 4.6. These six pairs of parameters, although independently measureable, are not independent of one another, as we shall now show. (Read the next paragraph while referring to Box 4.3.)

First, refer to the lines that start with Equilibrium exchange in Box 4.3. The two equilibrium exchange experiments must give the same values of $V_{\max}$ and K_m, since the flux in opposite directions, at any concentration, is the same, as we saw in Fig. 4.3. Thus $v_{io} = v_{oi}$. This reduces the number of $V_{\max}$, K_m pairs to five.

Second, consider what happens at low concentrations of substrate. For each procedure, the Michaelis–Menten equation [Eq. (4.1)] predicts that at low concentrations of substrate, the rate of transport should be given by $SV_{\max}/K_m$, where each $V_{\max}/K_m$ pair is in general different. [To show this, put S much less than K_m in Eq. (4.1). Then $v = SV_{\max}/K_m$.] At very low concentrations of substrate, however, the equilibrium exchange experiment in the *in-to-out* direction is the very same experiment as the zero-trans experiment in that direction, since the substrate concentration at *both* sides of the membrane is very low, and effectively zero. Thus $SV_{\max}/K_m$ for the equilibrium exchange experiment in the *in-to-out* direction must equal this term for the zero trans experiment in that direction. One can make exactly the same argument for the two experiments in the *out-to-in* direction. But the two equilibrium exchange experiments are identical, as we have just seen. We have, therefore, proved that all three values of $SV_{\max}/K_m$ *must* be equal to one another at the same value of S, and hence all three values of $V_{\max}/K_m$ must be identical. All three values are, indeed, the *permeability* of the system for that substrate, the ratio of the rate of transport to the substrate concentration (at low substrate concentration). Thus the permeability of the membrane for a substrate transported on a carrier is given by $V_{\max}/K_m$ for the equilibrium exchange and the two zero-trans experiments.

Third, consider the equilibrium exchange and infinite-trans experiments at very high ("infinite") concentrations of substrate on both sides of the membrane. The conditions here are the same for all these procedures since the substrate concentrations are infinite everywhere. Thus the maximum velocities of the infinite-trans and equilibrium exchange experiments must all equal one another, giving the results recorded in Box 4.3.

Table 4.1 (lower half) lists $V_{\max}$ and K_m values for the glucose transport system of the human red blood cell, obtained using these various experimental procedures.

4.6.2. Carrier Systems May Behave Asymmetrically

The asymmetric effects of the inhibitors described in Section 4.5 suggest that a carrier itself, as a protein embedded in the cell membrane, can be asymmetric. Indeed, the glucose carrier can behave asymmetrically in

Box 4.3. Kinetic Parameters for the Various Modes of Carrier Transport

We shall use the following symbols as super- or subscripts for the various kinetic parameters: superscript "zt" will describe a zero-trans experiment; superscript "ee" will describe an equilibrium exchange experiment; superscript "it" will describe an infinite-trans experiment; subscript "io" will describe an experiment in the in-to-out direction; and subscript "oi" will describe an experiment in the out-to-in direction.

Experiment	Direction	Transport velocity
Zero trans	In-to-out	$v_{io}^{zt} = S_i V_{io}^{zt}/(S_i + K_{io}^{zt})$
Zero trans	Out-to-in	$v_{oi}^{zt} = S_o V_{oi}^{zt}/(S_o + K_{oi}^{zt})$
Equilibrium exchange	In-to-out	$v_{io}^{ee} = S_i V_{io}^{ee}/(S_i + K_{io}^{ee})$
Equilibrium exchange	Out-to-in	$v_{oi}^{ee} = S_o V_{oi}^{ee}/(S_o + K_{oi}^{ee})$

but

$$v_{io}^{ee} = v_{oi}^{ee} \quad \text{(see text)}$$

therefore

$$V_{io}^{ee} = V_{oi}^{ee} = V^{ee} \quad \text{and} \quad K_{io}^{ee} = K_{oi}^{ee} = K^{ee}$$

These relationships are true for uncharged substrates. For charged substrates, the effect of the transmembrane potential has to be taken into account.

Experiment	Direction	Transport velocity
Infinite trans	In-to out	$v_{io}^{it} = S_i V_{io}^{it}/(S_i + K_{io}^{it})$
Infinite trans	Out-to-in	$v_{oi}^{it} = S_o V_{oi}^{it}/(S_o + K_{oi}^{it})$

but

$$V_{io}^{it} = V_{oi}^{it} = V^{ee} \quad \text{(see text)}$$

We are left with V_{io}^{zt}, V_{oi}^{zt}, V^{ee} and K_{io}^{zt}, K_{oi}^{zt}, K^{ee}, K_{io}^{it}, and K_{oi}^{it} and also the relations

$$V_{io}^{zt}/K_{io}^{zt} = V_{oi}^{zt}/K_{oi}^{zt} = V^{ee}/K^{ee} \quad \text{(see text)}$$

(Continued)

and

$$\text{Asymmetry} = K_{io}^{zt}/K_{oi}^{zt} = K_{io}^{it}/K_{oi}^{it} \text{ (see text, Section 4.10)}$$

This leaves only five independent parameters that have to describe six experimental protocols, in each of which both a term for V_{max} and one for K_m can be determined experimentally.

its transport behavior. Many experiments, especially at low temperatures or using outdated transfusion blood, have demonstrated that the K_m for glucose *entry* in zero-trans conditions can be quite a good deal smaller than its K_m for zero-trans *efflux* (see Table 4.1, lower part). There is nothing in the carrier model that would require a carrier to be symmetric, and, indeed, asymmetry is probably the rule rather than the exception. All that asymmetry requires is that certain of the rate constants of the model of Fig. 4.6 be different in the two directions across the membrane, a condition that is very often found. There are, however, some constraints on the asymmetry. Thus we have just shown that if the parameter K_m is asymmetric, the parameter V_{max} must be correspondingly asymmetric since V_{max}/K_m is identical for the two zero-trans procedures. In most cases asymmetry in a carrier arises from the situation that E_1 and E_2, the two conformations of the unloaded carrier, are of different free energy. Take the case (Fig. 4.6) where it is E_2 that is the more stable, i.e., that is of *lower* free energy. Then the free carrier will mostly be found in conformation E_2 and will be available *at* side 2 to transport substrate from that face. The fact that much carrier is available at side 2 means that V_{max} from that side is high—there is a good deal of it present to carry out transport. The K_m value is correspondingly high; one will need much substrate to saturate this abundance of carrier. In contrast, at side 1, carrier will be in short supply, V_{max} will be low from that side, and K_m will be correspondingly low, since only a low concentration of substrate is sufficient to saturate the small amount of carrier that is present at that face. The asymmetry can be defined as the ratio of the V_{max} terms *or* of the K_m terms. In the case where loaded carrier moves faster than unloaded carrier (so that the rate constants in g in Fig. 4.6 are higher than those in k), the asymmetry is determined by the ratio of the rate constants in k. This ratio is itself determined by the difference in free energy between the two conformations E_1 and E_2 of the unloaded carrier. We return to this important conclusion in Section 6.1.1 when we discuss the kinetic asymmetry of the sodium pump.

4.7. ELECTROGENIC ASPECTS OF CARRIER TRANSPORT

In this discussion, we have until now tacitly assumed that the substrate of the carrier was uncharged, a nonelectrolyte, such as the glucose molecule to which we have so frequently referred. But this assumption is not at all necessary and a number of important carrier systems are able to transport charged substrates, anions, or cations. Indeed, in Section 4.8 we shall go on to discuss the anion transporter of the red cell that moves bicarbonate and chloride ions across the cell membrane. Let us now consider the electrical aspects of the carrier transport of charged substrates.

In any transport system for which there is a net transport of charge across a membrane during a complete cycle of transport, there will be a mutual interplay between the rate of net transport and the electrical potential across the membrane. Such a transport system is termed *electrogenic*. Any individual step in the carrier model of Fig. 4.6 will be affected by the transmembrane potential, *if that step carries charge across the membrane*. It has, indeed, proved possible in a few cases to identify which step is affected by the potential. Let us, however, concentrate here rather on the effect of the transmembrane potential on the steady-state distribution of substrate. An important principle to grasp is that of the "complete cycle of transport." One has in mind here the complete set of events: a transporter loads up with substrate at the cis face of the membrane, reorients across the membrane, and delivers the substrate at the trans face. It returns to the cis face *either* loaded (when it has exchanged one substrate for another) or unloaded (when net transport has been brought about). If loaded, it unloads substrate once again, before completing the cycle. Only if a net charge is transferred across the membrane during an entire such cycle, will transport be electrogenic, i.e., will the membrane potential affect net transport and be affected by it. Consider the anion transporter of the red cell (to be discussed in detail in Section 4.8.1). It can carry a chloride ion across the membrane. If it comes back empty, it will have transferred one unit of negative charge across the membrane and it will therefore interact with the membrane potential. If it comes back loaded, either with chloride or with a bicarbonate ion, it will not. If it comes back loaded with a sulfate ion (which bears *two* negative charges), it will once again have performed an electrogenic event and will again interact with the membrane potential.

What is the form of this interaction with the membrane potential? We saw in Section 3.5 that an ion flowing across a membrane through a channel will set up a *diffusion potential* across that membrane. Nothing in our discussion at that point, however, depended on the route that the ion took through the membrane. In particular, whether an ion flows across the membrane through a channel, or on a carrier, or using some of the more complicated transport systems that we will discuss in Chapters 5 and 6, in all cases its flow across the membrane will set up a diffusion potential. Just as we saw in Section 3.5, the *size* of this membrane potential will depend on the fraction of the total transmembrane current that is carried by that ion; that is, it will depend on the concentration of the ion and its permeability (through whatever route it takes: channel, carrier, or other). In the event that there is no other route for current flow, that is, the ion carrier carries *all* the current that is flowing across the membrane, then, as we saw in Section 3.5, the diffusion potential ($\Delta\psi$) will be determined by the transmembrane concentration gradient (S_{II}/S_I) and the charge on the substrate, Z, according to the Nernst or equilibrium potential equation [Eq. (2.5)], which we reproduce here:

$$\Delta\psi = RT/ZF \ \ln(S_{II}/S_I) \tag{2.5}$$

This equation is as valid for the movement of an ion via a carrier as for its movement through a channel. Net flow of the ion via the carrier will cease when the membrane potential and the concentration gradient of the substrate are related as in Eq. (2.5). (This equation is, of course, valid also for uncharged substrates for which the charge Z is equal to zero, when the ratio of substrate concentrations is unity for all values of the membrane potential.)

There are three ways in which the effect of an electrogenic transport system can be detected experimentally. The first is applicable when there is no major route for current flow across the membrane other than ion flow on the carrier itself. A net flow of substrate will occur until the membrane develops the potential given by Eq. (2.5). This potential can be measured by any of the methods discussed in Section 3.4.2. Evidence for the development of such a potential during substrate flow is thus evidence that the transport system is electrogenic. In the second method, when a potential is *imposed* across the membrane by means of an externally applied electrode system or by an imposed and overwhelming diffusion potential, a measurable net flow of substrate will occur in accordance with the prevailing transmembrane potential, until the substrate concentration ratio is that given by Eq. (2.5). Finally, in the technique known as *short-circuiting*, the transmembrane potential is maintained at zero by connecting up the solutions bathing the two faces of the membrane by a current-carrying wire. Substrate moving down its concentration gradient, on the carrier, will carry current across the membrane. This current will flow back through the wire connecting up the two faces of the membrane, completing the electrical circuit. The strength of this current flow can be measured by an external ammeter and is a direct measure of the rate of net movement of the transported substrate, i.e., of the rate of electrogenic carrier-mediated transport. Note that we can "short-circuit" the membrane in another way. If we add to the system a channel or carrier for the ion that the original carrier in the membrane is transporting (i.e., if we add one of the ionophores such as gramicidin or valinomycin), we have, in effect, replaced the wire in the previous description by the ionophore. The current carried across the membrane on the carrier of the membrane can flow back through the ionophore. If sufficient ionophore is present, the transmembrane potential will remain at zero and transport on the electrogenic carrier will continue until the transmembrane concentration gradient of the substrate is equal to unity.

4.8. SOME INDIVIDUAL TRANSPORT SYSTEMS

We used data on the glucose transporter of the human red blood cell to illustrate the properties of carrier systems in general. Although we know so much about the properties of the system, we do not know what its role in the life of the cell is. Its rate of operation is orders of magnitude faster than the rate of consumption of glucose by the cell. One possible role that has been suggested is that the system allows the plasma glucose to come into rapid equilibrium with the red cell interior, thus, in effect, expanding the glucose-carrying capacity of the blood. There are many other carriers, however, whose biological role is more established and we can learn a good deal by studying them. We shall consider three systems in greater detail.

4.8.1. The Anion Transport System: Band 3

The primary function of the red blood cell is to carry oxygen from the lungs to the tissues, the oxygen being bound inside the cell to hemoglobin. A secondary role of this cell is to carry carbon dioxide from the tissues to the lungs. Carbon dioxide dissolves in water to form carbonic acid. Thus what the red cell actually carries is carbonic acid. (Box 4.4 discusses carbon dioxide transport in the blood in greater detail). The protons of the acid are bound to hemoglobin, and the bicarbonate is carried dissolved in the cell water. Its entry into and exit from, the cell is, however, dependent on the action of a carrier which *exchanges* bicarbonate ions for chloride ions. So important is this role of bicarbonate transport that the bicarbonate–chloride ion carrier or anion carrier turns out to be the most prevalent

Box 4.4. The Physiologic Role of the Anion Transporter of the Red Cell

The physiologic role of the anion transporter is to exchange chloride ions for bicarbonate ions across the red blood cell membrane, thereby considerably increasing the carbon dioxide-carrying capacity of the blood. Overall, the system seems to work like this: CO_2 is, of course, produced in the tissues during cellular oxidation reactions, and this CO_2 has to be transported in the blood to the lungs in order to be expelled from the body. Movement of carbon dioxide is in the direction of its concentration gradient (higher in the tissues than in the atmosphere), but the rate of this movement will be governed by the *resistances* of the various barriers that have to be overcome. One such barrier is presented by the limited solubility in blood of CO_2 as CO_2, which limits the carrying capacity of the blood. Chemical reaction of CO_2 with water to give carbonic acid (H_2CO_3) allows more CO_2 to dissolve, forming HCO_3^- ions and H^+, the latter being titrated by cell and blood buffers. The chemical reaction of CO_2 with H_2O is catalyzed by the enzyme carbonate dehydratase, found within the red cell. (If carbonate dehydratase is inhibited, hydration of the CO_2 is reduced by many orders of magnitude.) For the CO_2 hydration reaction to continue, the products of reaction, HCO_3 and H^+, must disappear. The H^+ is removed by being bound to hemoglobin while the bicarbonate ion leaves the cell, in exchange for chloride ion, by the anion transport system. In the lungs this whole reaction sequence is reversed. Any bicarbonate that enters the red cell in exchange for chloride combines with protons driven off from hemoglobin as oxygen binds to it. The resulting carbonic acid is dehydrated to CO_2, which exceeds its solubility limit within the blood and is expelled into the atmosphere.

A detailed kinetic analysis of the rates of transport of bicarbonate and of chloride at body temperature enabled Jens Wieth and colleagues to quantitate the importance of anion transport in human physiology. Knowing the physiologically relevant values of the chloride and bicarbonate ion concentrations at the two membrane faces of the red cell and the actual measured transport capacity of the cell for these anions, they calculated that exchange of chloride for bicarbonate takes about 0.4 sec to reach 90% completion. The time that a red cell takes to pass through the lung capillaries is about 0.7 sec when the body is at rest and reduces to 0.3 sec during periods of maximum work, when cardiac output is greatest. Clearly, during periods of active stress, anion trans-

(Continued)

port is a limiting factor in the capacity of blood to carry CO_2 for expulsion. This is despite the facts that the anion transport system is enormously abundant in the red cell, and that the system itself is acting very rapidly, having the high turnover number of about 10^4 exchanges per second at room temperature and, presumably, an even higher number at body temperature.

protein of the red cell membrane, being present at more than a million copies per cell. (In contrast, there are about 300,000 copies of the glucose carrier, 12,000 copies of a uridine carrier, and only 1000 copies of the sodium pump, which we discuss in Section 6.1.) The importance of the anion carrier to the economy of the cell and its abundance have both prompted much research on this system, which is one of the best known membrane carriers.

During the search for a chemical reagent that would label the anion carrier and allow its identification in the membrane, Ioav Cabantchik and Aser Rothstein found that derivatives of stilbene (1,2-diphenylethylene) inhibited anion transport in rather a specific manner. One of these, 4,4'-diisothiocyano-2,2'-stilbenedisulfonic acid (DIDS) (see Fig. 4.9), is a widely used reagent, giving excellent inhibition of anion transport, combining with the anion transporter to form a stable derivative. When isotopically labeled DIDS is reacted with red cells, a single labeled protein is observed when the proteins of the cell membrane are solubilized in detergent and separated electrophoretically (Fig. 4.9). The label is present in the third fastest-moving major band of protein, and the name "band 3" has remained as the nickname of the anion transporter.

The band 3 protein has a molecular mass of 97,000 Da. It is a typical *intrinsic* membrane protein (see Section 1.2), being firmly bound to the membrane and released only by treating the membranes with detergents.

When intact red cells are treated with the proteolytic enzyme chymotrypsin, from the outside of the cell, the polypeptide chain of band 3 is split at a single point to yield two chains of mass 55,000 and 42,000 Da, termed the "AB" and "C" portions, respectively. Such a split does not lead to loss of function. If the enzyme trypsin is allowed to enter the red cell and is present only at the cytoplasmic face of the membrane, it can attack band 3, splitting off a polypeptide of 38,000-Da molecular mass (termed "A") and again leading to no loss of transport activity. If the chymotrypsin treatment is carried out first (yielding AB and C), trypsin splits AB into A and B, giving rise to three polypeptides in all. Clearly,

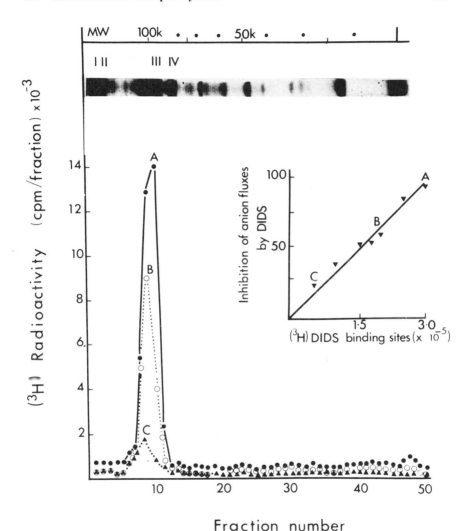

Fraction number

Fig. 4.9. Electrophoresis of red cell membrane proteins and of band III. The figure shows (top) a Coomassie-blue-stained polyacrylamide gel of the total membrane proteins of the human red blood cell, solubilized and electrophoresed in sodium dodecyl sulfate. Above the gel are indicated the relative molecular masses of the different polypeptides numbered I, II, III, and IV according to the convention. The red cells had been treated for three different time periods (A, B, and C) with ^{3}H-labeled DIDS (see text). The gel was cut into strips and the radioactivity in each strip measured. The resulting counts are shown in the lower half of the figure. Almost all the radioactivity is associated with band III. The inset shows a plot for experiments done at a number of different times, where the degree of inhibition of anion transport is plotted on the ordinate against counts per minute associated with band III, on the abscissa; A, B, and C in the inset refer to the preparations shown in the main figure. (Original figure, reproduced with kind permission of Z. I. Cabantchik.)

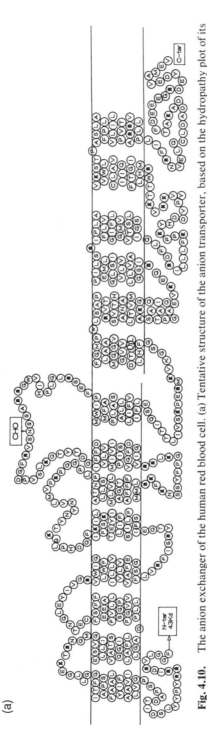

(a)

Fig. 4.10. The anion exchanger of the human red blood cell. (a) Tentative structure of the anion transporter, based on the hydropathy plot of its amino acid sequence. The letters refer to the amino acids in their single-letter coding. CHO is the point of attachment of the glycosyl side chain. The 14 putative membrane-spanning helices are shown as embedded in the bilayer. (b) Cutaway model of a possible transmembrane disposition of the membrane-spanning helices. Numbers refer to respective helices in Fig. 4.10(a). Lysine groups at the active center are depicted as filled circles; the filled square is an arginine residue, and the hatched circle is a glutamate. Fig. 4.10(a) and (b) both drawn by Z. I. Cabantchik and O. Eidelman. (*Figure continues*)

(b)

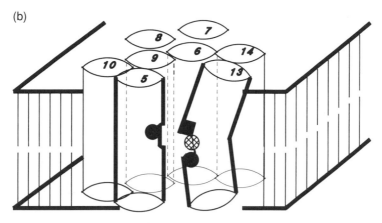

Fig. 4.10 (*Continued*)

one end of peptide B is available to chymotrypsin at the extracellular face of the membrane, the other to trypsin at the cytoplasmic face. Peptide B must therefore *span* the membrane. Indeed, similar pieces of evidence lead to the conclusion that band 3 spans the membrane very frequently.

The gene coding for the anion transporter has been cloned and sequenced and the amino acid sequence of the protein deduced [Fig. 4.10(a)]. The hydropathy plot of this sequence leads to the tentative structure depicted in Fig. 4.10(a), which is fully in agreement with the chemical evidence regarding the disposition of the protein across the membrane. A number of regions of the protein have been identified as being at or near the active center of the carrier. The region that binds DIDS, for instance, is one such region. Figure 4.10(b) depicts a possible three-dimensional "structure" for the molecule, seen as a cutaway model. The extracellular surface of the protein is the uppermost surface in the figure. The numbers 5–10, 13, and 14 refer to the membrane-spanning α helices numbered from the N-terminal end of the protein. This depiction is not totally fanciful. Cabantchik and colleagues have studied the inhibitory potencies of a large number of chemical derivatives of DIDS, including those in which the active group in the inhibitor is separated from a bulky "tail" by a shorter or longer chain of carbon atoms. By this means, they were able to show that the DIDS-binding residue on the protein is situated some depth into the molecule itself, so that the DIDS-binding center [depicted as the uppermost filled circle in Fig. 4.10(b)] is in a wide-mouthed cleft of the transporter.

The anion transporter has a very wide specificity for its substrates. Bicarbonate and chloride ions, the natural substrates, are themselves very different in chemical structure, but the range of substrates that the anion transporter moves can be extended to include some of the common inorganic anions (nitrate, thiocyanate, sulfate) and many organic anions, including such large ions as malonate and phthalate. It is of interest that many of these anions have much the same "affinity" for the anion carrier (measured as K_m), but have very different maximal velocities. For instance, V_{max} for sulfate transport is about 10,000 times slower than that of chloride, whereas the K_m parameters are 4.4 mM for sulfate and 3.9 mM for chloride. This behavior is quite different from that found for most other carriers, with V_{max} generally of much the same value for the different substrates that a particular carrier can transport, but K_m values varying over a wide range (see Table 4.1, upper part, for the glucose transporter). These facts probably tell us something important about the anion transporter, but we do not yet know what this is.

The anion transporter shows very clear evidence for *recruitment* (as discussed for the glucose carrier in Section 4.5). Nonpenetrating analogs of the inhibitor DIDS, applied extracellularly, react much more rapidly with the anion transporter when chloride ions are present inside the cell and sulfate outside than with these ions in the reverse orientation. We have just seen that sulfate crosses the membrane on the carrier far more slowly than does chloride. Thus with chloride outside and sulfate inside, most of the carrier is shunted to the cytoplasmic face of the membrane, where no inhibitor is present. The rate of inhibition is slow. In the reverse orientation of these ions, the carrier is recruited to the extracellular face of the membrane where it can react with the DIDS analog and be inhibited.

There is good evidence that a glutamate residue is alternately available for reaction with a chemical reagent (the so-called Woodward's reagent K, or N-ethyl-5-phenylisoxazolium 3′-sulfonate) first at one face and then at the other face of the membrane. This change of availability of the glutamate to the chemical reagent is brought about during a transport event. In Fig. 4.10(b) the hatched circle is meant to depict this glutamate residue, and it is a change in the disposition of the transmembrane helices that is thought to allow the glutamate to reorient with respect to the plane of the membrane.

We have seen that the function of the anion carrier is to *exchange* chloride with bicarbonate ions. It is very ineffective at bringing about a *net* transport of either of these ions. Indeed, exchange of chloride for chloride (measured in an equilibrium exchange experiment as in Section 4.2.3) is some *four orders of magnitude* faster than its net transport (measured in a zero trans experiment as in Section 4.2.1). On the model of Fig. 4.6, this means that the anion-loaded carrier reorients 10^4 times faster than does the unloaded carrier. There is a very good physiological role for this great reduction in net transport of anions. Were net transport of chloride to be fast in the face of reasonably high rates of potassium movement through the potassium-specific channels of the red cell, net movement of potassium chloride would be allowed and the cells would lose KCl continually. [The Donnan potential (Section 3.7) across the red cell membrane is low, and the concentration of potassium ions is maintained high in this cell by ion pumping.] Evolving an exchange-only anion transporter has meant that the red cell can transport bicarbonate ions into and out of the cell, in exchange for chloride ions, without a net movement of KCl.

4.8.2. The Amino Acid Carriers

In contrast to the glucose carrier (which transports a wide variety of sugars) and the anion transporter (whose specificity for different anions is, as we have just seen, very broad), the amino acid carriers have a narrower, *class* specificity that has been delineated by many years of patient work. Table 4.3 records the specificities of seven different amino acid transport systems found in animal cells. These are of great importance in allowing amino acids to enter these cells and take part in protein synthesis. (Bacterial cells have many more such systems, often of an even narrower specificity.) Note from Table 4.3 the y^+ system, which transports basic amino acids (arginine and lysine), and the x^- system, which moves the acidic amino acids (glutamate and aspartate). The remaining five systems transport neutral amino acids of particular groups. The specificity of the various systems is such that many systems, acting in parallel, may be involved in the uptake of a single amino acid.

TABLE 4.3

Classification of the Amino Acid Transporters[a]

Name of system	Preferred substrates	Sodium requirement[b]
A	Alanine, aminoisobutyric	Yes
ASC	Alanine, serine, cysteine, threonine	Yes
Gly	Glycine, sarcosine	Yes
L	Leucine, phenylalanine	No
N	Glutamine, histidine, asparagine	Yes
y^+	Arginine, lysine, ornithine	No
x^-	Glutamate, aspartate (as anions)	Yes

[a] Based largely on the work of Christensen (1984).
[b] For the importance of the sodium requirement of these systems, see Chapter 5.

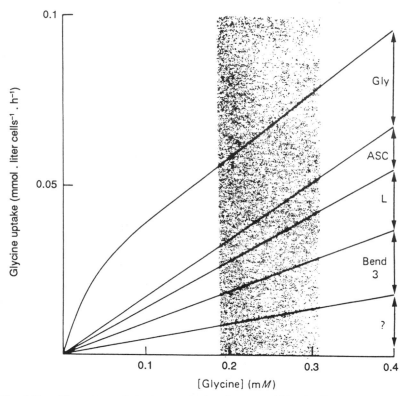

Fig. 4.11. Five routes of glycine uptake into human red blood cells. The upper curve shows total glycine uptake as a function of glycine concentration (at 37°C). The routes marked Gly, ASC, and L represent uptake by three specific amino acid transporters, so designated. The route marked "Bend 3" designates glycine that is transported on the anion transporter (band 3), while the route marked with a question mark (?) is transport by a route that is still to be identified. Taken, with kind permission, from J. C. Ellory, S. E. M. Jones, and J. D. Young (1981). *J. Physiol.* (*London*) **320**, 403–422.

Figure 4.11 shows uptake by human red cells of the amino acid glycine, as a function of its concentration. (Refer to Table 4.3 for the definition of the various systems.) The curves, from highest to lowest, show successively (1) the total uptake, (2) uptake when a specific inhibitor of the Gly carrier is added, (3) uptake when this inhibitor, together with an inhibitor of the ASC carrier, is added, (4) uptake when in addition the L system is inhibited, and (5) uptake when the anion transporter (which can also transport glycine) is, in addition, blocked. Even after all these glycine-carrying systems are blocked by specific inhibitors, there remains a small but significant component that has no known inhibitor (the bottom curve in Fig. 4.11).

Many of the amino acid transporters can bring about the concentration of their substrates by the process of cotransport with sodium, which we will discuss in detail in Chapter 5.

4.8.3. The Choline Transporter of the Human Red Blood Cell

Choline is a very important component of many of the phospholipid molecules that make up the membranes of cells. Carrying a strong positive charge and hence being very lipid-insoluble, choline itself cannot cross cell membranes at a significant rate by simple diffusion, nor is the choline molecule small enough to be able to cross through the ion-specific channels. Yet choline must be enabled to enter cells in order to be harnessed to phospholipid synthesis. Very detailed kinetic studies have been performed on a system in human red cells that moves choline across the membrane of this cell. The system shows the features of recruitment that we have already discussed, and much solid evidence has been accumulated that the carrier exists in two conformations that interorient during the transport event. The system is specific for choline and its derivatives, but a wide range of affinities and transport velocities prevails within this grouping. (Table 4.4 lists some transport parameters that have been derived for this system for different substrates.) Devés and Krupka pointed

TABLE 4.4

Choline Transporter of the Human Red Blood Cell: Affinities and Rates for Substrate Analogs[a]

Substituent R group	K_m (μM)	V_{max} (relative to choline)
Methyl (i.e., choline proper)	6	1
Ethyl	12	0.93
n-Propyl	33	0.68
Isopropyl	303	0.25
n-Butyl	30	0.06
n-Pentyl	7.3	0.02
n-Decyl	0.3	0.006
n-Dodecyl	0.095	0.01

Core structure is:

$$R\text{-}^+N\text{-}CH_2CH_2OH$$

with CH_3 and CH_3 substituents.

[a] Data from R. Devés and R. M. Krupka, *Biochim. Biophys. Acta* **557,** 469–485 (1979).

out that the affinities and maximum velocities do not in any way correlate with one another. A substrate bearing a long hydrophobic tail can demonstrate a very high affinity for the carrier. Yet the long-chain derivatives give very low values for the rates of conformation change of the loaded forms of the carrier. It is as if the long chains enable the substrate to bind very effectively to (or near) the active center of the carrier, but do not allow that reorientation of the carrier–substrate complex that constitutes the transport event. Figure 4.12 is an attempt to make this point visually.

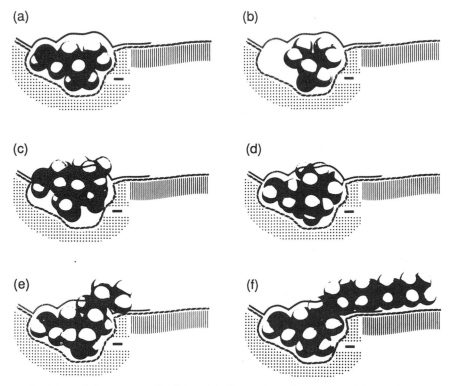

Fig. 4.12. Substrates and inhibitors binding at the active center of the choline transporter. The molecular models show four different substrates [(a)–(d)] and two inhibitors [(e), (f)] of the choline transporter more or less well accommodated by its binding site. Substrate (a) is choline itself, which fits the site well, has a good affinity, and is well transported. Substrate (b) is tetramethylammonium, which is rather small to bind well to the site, but is transported readily. A choline analog such as that depicted in (f) binds exceedingly well to the site but is too bulky to be transported. Taken, with kind permission, from R. Devés and R. M. Krupka (1979). *Biochim. Biophys. Acta* **557**, 469–485, for Dept. of Agriculture, Govt. of Canada.

4.9. AN OVERALL VIEW OF THE MEMBRANE CARRIERS: CARRIERS AND CHANNELS COMPARED

We have discussed in some detail a number of carrier systems, all, as it happens, from human red blood cells. It must be stressed, however, that *all* cells probably contain membrane carriers and in most cases a bewildering variety of carriers exist in the same cell membrane. They can carry most of the common metabolites ranging from simple ions through to complex vitamins and nucleotides. All the membrane-residing carrier systems that have been studied in detail present the following common features: (1) they are membrane-bound intrinsic proteins that span the cell membrane; (2) they show specificity for their substrates, although this may be narrow or broad; (3) they can be inhibited by specific inhibitors that can be protein reagents (when the inhibition is usually noncompetitive) or substrate analogs (when competition is the rule); (4) they show saturation, that is, they display Michaelis–Menten kinetics (although sometimes of a complex type); (5) they show evidence of existing in two conformations, a reorientation step allowing the substrate to have access to the binding sites of the carrier, first at one side of the membrane and then at the other. It remains for researchers to elucidate the details of these reorientation steps and to establish how binding of substrate to carrier allows the reorientation to take place, often at a rate far faster than the corresponding step in the absence of substrate binding.

Working out the details of the transport event will probably require the crystallization of the membrane protein and the determination of its structure in its two conformations, using X-ray crystallographic methods. As an indication of how much detail can be found by such studies consider Fig. 4.13, which shows the striking difference between two conformations of the (water-soluble) enzyme hexokinase when it is not bound to substrate [Fig. 4.13(a)] or when loaded with glucose [Fig. 4.13(b)]. This is how far the X-ray crystallographic analysis of soluble proteins has advanced. The fact that the membrane carriers are embedded in a lipid matrix has prevented the development of routine procedures for their crystallization in three dimensions. (But see Section 1.4 for a discussion of the progress that has been achieved already!) It seems that only the X-ray analysis of such crystals will yield structural knowledge with the resolution of Fig. 4.13. We must reach the same degree of certainty before we can form detailed models of the events in carrier transport.

Why should cells have evolved both carriers and channels? What is so special about carrier functions that channels cannot accomplish? We saw that the channels have only a limited specificity, depending largely on

(a)   (b)

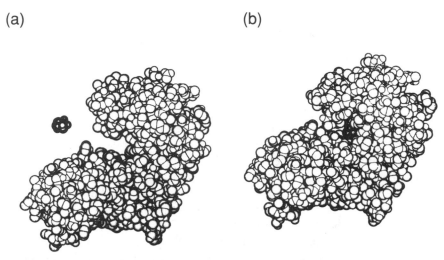

Fig. 4.13.   The conformation change when an enzyme binds its substrate. Computer-simulated diagram of the enzyme hexokinase before (a) and after (b) binding to its substrate glucose. Note how the "jaw" of the enzyme has closed about the glucose molecule. This might be a model of how a transporter's conformation is altered when it binds its substrate. (Original figure, courtesy of T. A. Steitz.)

their ability to act as size-discriminating filters. There has been no evolution of a channel that can discriminate between sugars of the same overall size, or which can allow large amino acids to pass, but not those of smaller size. Yet just these properties are easily within the power of the carriers, with their ability to bind specific substrates with precision, preferring one sugar to another or picking out one amino acid among many. This pedantic specificity is attained, however, at the expense of a slow *rate* of transport. The maximum velocity of action of all carriers is limited by the rate of the conformation changes that they undergo (see Table 4.2). In other proteins, these rates are seldom above 1×10^6 sec^{-1}, and are more often in the 1000 sec^{-1} range. Carriers, too, have maximum velocities (per molecule of carrier) in the range of tens to thousands of transport events per second. In contrast, as we saw in Section 3.1, a channel can transport tens of millions of its substrate ions per second. It seems that carriers are the transport system of choice when discrimination between large substrates is required. The loss in speed that this discrimination brings with it is compensated for by the synthesis of many copies of the carrier by the cell and their insertion in the cell membrane. When small ions are to be transported and high rates of transport are required, channels seem to be used in preference.

4.10.   THE FULL EQUATION FOR CARRIER TRANSPORT

Most students will not find it necessary to go into any more detail into transport kinetics at this point, but for those who do, the full equation that describes the carrier model of Fig. 4.6 is given in Box 4.5, together with the interpretation of the various experimentally determinable parameters in terms of the rate constants of the model.

What is recorded is an equation that describes how the *unidirectional flux* of a substrate (which we write as $v_{1\to2}$, for the flux in the direction 1 to 2) depends on the substrate concentrations at the two sides of the membrane. The unidirectional flux can be measured as the rate of movement of isotope labeled substrate, as we described above in Section 4.2.3 for the equilibrium exchange and in Section 4.2.4 for the infinite-trans experiments. Any *net flux* is equal to the difference between the unidirectional fluxes in the two directions. Thus if we have an equation for the unidirectional fluxes, we can obtain the net flux and hence describe *any* transport experiment.

It turns out that any unidirectional flux can be described in terms of four *membrane resistances* and an affinity, as defined in Box 4.5. The solution is given in Stein (1986). The four resistances are related by a simple equation (Table 4.5), so only three need to be measured independently. Thus only four parameters, an affinity and three of the four resistances, are required to define any transport system that behaves according to the model of

---

Box 4.5. Steady-State Solutions for the Simple Carrier

Analysing Fig. 4.6 we show that

$$v_{1\to2} = \frac{(K + S_2)S_1}{K^2 R_{00} + KR_{12}S_1 + KR_{21}S_2 + R_{ee}S_1S_2}$$

$$nR_{12} = \frac{1}{b_2} + \frac{1}{k_2} + \frac{1}{g_1} + \frac{g_2}{b_2 g_1}$$

$$nR_{21} = \frac{1}{b_1} + \frac{1}{k_1} + \frac{1}{g_2} + \frac{g_1}{b_1 g_2}$$

$$nR_{ee} = \frac{1}{b_1} + \frac{1}{b_2} + \frac{1}{g_1} + \frac{1}{g_2} + \frac{g_1}{b_1 g_2} + \frac{g_2}{b_2 g_1}$$

$$nR_{oo} = \frac{1}{k_1} + \frac{1}{k_2}$$

$$K = \frac{k_1}{f_1} + \frac{k_2}{f_2} + \frac{b_1 k_1}{f_1 g_1}$$

Constraint: $b_1 f_2 g_2 k_1 = b_2 f_1 g_1 k_2$

These results are for the special case of an uncharged (neutral) permeant (Stein, 1986). Flux $v_{1\to2}$ is the unidirectional flux from solution 1 to solution 2; the reverse flux $v_{2\to1}$ can be obtained by interchanging subscripts 1 and 2 in the equation for $v_{1\to2}$; S_1 and S_2 are the permeant concentrations in solutions 1 and 2, while n is the total concentration of carriers.

TABLE 4.5

Interpretation of Experimental Data in Terms of the Basic Observable Parameters for the Simple Carrier

Procedure	V_{max}	K_m	$\pi(= \lim\limits_{S \to 0} v/S)$
Zero trans	$V_{12}^{zt} = \dfrac{1}{R_{12}}$	$K_{12}^{zt} = K\dfrac{R_{00}}{R_{12}}$	$\pi_{12}^{zt} = 1/(KR_{00})$
	$V_{21}^{zt} = \dfrac{1}{R_{21}}$	$K_{21}^{zt} = K\dfrac{R_{00}}{R_{21}}$	$\pi_{21}^{zt} = 1/(KR_{00})$
Infinite trans	$V^{it} = V_{12}^{it}$	$K_{12}^{it} = K\dfrac{R_{21}}{R_{ee}}$	$\pi_{12}^{it} = 1/(KR_{21})$
	$= V_{21}^{it} = \dfrac{1}{R_{ee}}$	$K_{21}^{it} = K\dfrac{R_{12}}{R_{ee}}$	$\pi_{21}^{it} = 1/(KR_{12})$
Infinite cis	$V_{12}^{ic} = \dfrac{1}{R_{12}}$	$K_{12}^{ic} = K\dfrac{R_{12}}{R_{ee}}$	—
	$V_{21}^{ic} = \dfrac{1}{R_{21}}$	$K_{21}^{ic} = K\dfrac{R_{21}}{R_{ee}}$	—
Equilibrium exchange	$V^{ee} = \dfrac{1}{R_{ee}}$	$K^{ee} = K\dfrac{R_{00}}{R_{ee}}$	$\pi^{ee} = 1/(KR_{00})$

Constraint: $R_{ee} + R_{00} = R_{12} + R_{21}$

the simple carrier. The full equation that describes any unidirectional flux is

$$v_{1 \to 2} = \frac{KS_1 + S_1S_2}{K^2R_{oo} + KR_{12}S_1 + KR_{21}S_2 + R_{ee}S_1S_2} \tag{4.3}$$

where the terms in R and K are as defined in Box 4.5. The unidirectional flux in the opposite direction $(2 \to 1)$ is given simply by exchanging the numbers 1 and 2 throughout Eq. (4.3). The net transport of the substrate S is given by the difference between the two forms of Eq. (4.3), that given and that with numbers 1 and 2 interchanged throughout. Note that the net flux so derived has a term $S_1 - S_2$ in its numerator. Thus when $S_1 = S_2$ the net flux disappears, showing that the simple carrier of Fig. 4.6 cannot, of course, pump substrate up a concentration gradient. Equation (4.3) looks formidable but is very easy to manipulate. If we want to use it to find the equation that describes the zero-trans flux in the $1 \to 2$ direction, we merely put S_2 equal to zero in Eq. (4.3), and after simplification we obtain

$$v_{1 \to 2} = S_1/(KR_{oo} + R_{12}S_1) \tag{4.4}$$

This is of the exact form of the Michaelis–Menten equation [Eq. (4.1)], if we realize that $V_{max} = 1/R_{oo}$ and $K_m = KR_{oo}/R_{12}$. The maximum velocities of the equilibrium exchange experiment and of the infinite-trans experiment can be found similarly by putting $S_1 = S_2 =$

S in Eq. (4.3) for the former and S_2 equal to infinity for the latter. The results thus obtained are collected in Table 4.5.

Why do we call the terms in R "resistances" and K an "affinity"? Note that all the maximum velocities are given in Table 4.5 as reciprocals of a resistance parameter. Maximum velocities are maximal conductances of the membrane for the substrate in question, and hence their reciprocals are properly called *resistances*. All the Michaelis constants are found in Table 4.5 as the product of K and a ratio of resistances. Hence K has the dimensions of a half-saturation concentration. It measures, indeed, the "virtual affinity" of carrier for substrate, the affinity when the equilibrium between the two conformations of the carrier is undisturbed by any substrate that is present.

Note that, in principle, each different type of experiment (equilibrium exchange, zero trans, infinite trans) leads to a different V_{max} and a different K_m. There is no such thing as a K_m for a transport system as such, if the particular type of experiment performed is not specified. Note, too, that each of the K_m parameters contains K, which is the same for all. Thus the difference between them arises from the resistance terms. One should verify from Table 4.5 that if the conformation changes of the loaded carrier are faster than those of the free carrier (i.e., if rate constants in g are faster than those in k), the K_m for a zero-trans experiment will be smaller than that for the equilibrium exchange experiment and, again, smaller than that for the infinite-trans experiment. This result has a simple intuitive basis. If the free-carrier conformation changes are slow, most of the carrier in a zero trans experiment ends up at the trans face, waiting for the conformational change that will bring its binding sites back to the cis side from which transport is taking place. Very little carrier will be available at cis face for binding to the substrate. Thus the maximum velocity of transport is slow and a low concentration of substrate is sufficient to bind enough carrier to reach half this maximum velocity.

In contrast, in an infinite-trans experiment, all the carrier is at the cis face, since any that reaches the trans face is immediately bound to substrate there and a rapid conformation change reorients it so as to face the cis side. The maximum velocity of transport is higher than in the zero-trans experiments in either direction (all carrier is ready for this cis $\rightarrow$ trans transport) and a high concentration of substrate is required to half-saturate this bumper harvest of carrier.

The asymmetry of the transport behavior of the carrier is also easily understood from Eq. (4.3) and Table 4.5. Note, first, that the definitions of the parameters K, R_{oo} and R_{ee} are unchanged when one interchanges the numbers 1 and 2 in Box 4.5. Thus, these parameters behave symmetrically. The only place where asymmetry can enter is in the remaining resistance terms, R_{12} and R_{21}. Indeed, the asymmetry of a transport system can be usefully defined by the ratio R_{12}/R_{21}. This ratio then defines (see Table 4.5) the ratio of maximum velocities in the zero-trans experiments and the ratio of K_m terms in the zero and infinite-trans experiments.

SUGGESTED READINGS

General

Jarvis, S. M. (1989). Uniport carriers for metabolites. *Curr. Opin. Cell Biol.* **1**, 721–728.
Stein, W. D. (1986). "Transport and Diffusion across Cell Membranes," Chapter 4. Academic Press, Orlando, Florida.

Inhibition

Basketter, D. A., and Widdas, W. F. (1978). Asymmetry of the hexose transfer system in human erythrocytes. Comparison of the effects of cytochalasin B, phloretin and maltose as competitive inhibitors. *J. Physiol. (London)* **278**, 389–401.
Devés, R., and Krupka, R. M. (1978). Cytochalasin B and the kinetics of inhibition of biological transport. A case of asymmetric binding to the glucose carrier. *Biochem. Biophys. Acta* **510**, 339–348.
Devés, R., and Krupka, R. M. (1980). Testing transport systems for competition between pairs of reversible inhibitors. Inhibition of erythrocyte glucose transport by cytochalasin B and steroids. *J. Biol. Chem.* **255**, 11870–11874.

Molecular Biology of Glucose Transporter

Blok, J. *et al.* (1988). Insulin-induced translocation of glucose transporters from post-Golgi compartments to the plasma membrane of 3T3-L1 adipocytes. *J. Cell Biol.* **106**, 69–76.
James, D. E. *et al.* (1989). Molecular cloning and characterization of an insulin-regulatable glucose transporter. *Nature (London)* **338**, 83–87.
Mueckler, M. *et al.* (1985). Sequence and structure of a human glucose transporter. *Science* **229**, 941–945.
Thorens, B. *et al.* (1988). Cloning and functional expression in bacteria of a novel glucose transporter present in liver, intestine, kidney, and beta-pancreatic islet cells. *Cell (Cambridge, Mass.)* **55**, 281–290.

Kinetics of Carrier Transport

Lieb, W. R., and Stein, W. D. (1974). Testing and characterising the simple carrier. *Biochem. Biophys. Acta* **373**, 178–196.
Wheeler, T. J., and Whelan, J. D. (1988). Infinite-cis kinetics support the carrier model for erythrocyte glucose transport. *Biochemistry* **27**, 1441–1450.

Valinomycin

Benz, R., and Läuger, P. (1976). Kinetic analysis of carrier-mediated ion transport by the charge-pulse technique. *J. Membr. Biol.* **27**, 171–191.
Stark, G. *et al.* (1971). The rate constants of valinomycin-mediated ion transport through thin lipid membranes. *Biophys. J.* **11**, 981–994.

Two Conformations of the Carrier

Krupka, R. M. (1985). Reaction of the glucose carrier of erythrocytes with sodium tetrathionate: Evidence for inward-facing and outward-facing carrier conformations. *J. Membr. Biol.* **84**, 35–43.
Krupka, R. M., and Devés, R. (1983). Kinetics of inhibition of transport systems. *Int. Rev. Cytol.* **84**, 303–352.

Electrogenic Aspects of Carrier Transport

Blaustein, M. P., and Lieberman, M., eds. (1984). "Electrogenic Transport: Fundamental Principles and Physiological Implications." Raven Press, New York.

Anion Transporter

Cabantchik, Z. I., and Eidelman, O. (1990). The anion transporter of the human red blood cell. *Subcellular Biochemistry* (in press).

Kopito, R. R., and Lodish, H. F. (1985). Primary structure and transmembrane orientation of the murine anion exchange protein. *Nature (London)* **316,** 234–238.

Wieth, J. O. *et al.* (1982). Chloride-bicarbonate exchange in red blood cells: Physiology of transport and chemical modification of binding sites. *Philos. Trans. R. Soc. London, B Ser.* **299,** 383–399.

Amino Acid Transporters

Christensen, H. N. (1984). Organic ion transport during seven decades. The amino acids. *Biochim. Biophys. Acta* **779,** 255–269.

Choline Transport

Devés, R., and Krupka, R. M. (1984). The comparative specificity of the inner and outer substrate transfer sites in the choline carrier of human erythrocytes. *J. Membr. Biol.* **80,** 71–80.

CHAPTER 5

Coupling of Flows of Substrates: Antiporters and Symporters

All the modes of transport that we have considered up to now act so as to move their substrates *down* their electrochemical gradients (defined in Section 2.2). Whether transport is by simple diffusion, by diffusion through channels, or by diffusion on the simple carrier, the laws of such transport require that the rate of transport depends (for uncharged substrates) on the concentration difference $S_1 - S_2$ so that at equilibrium, when net movement ceases, $S_1 = S_2$ and the concentration gradient disappears. {Recall from Section 2.2 that, for a charged substrate, net movement ceases when $S_1 = S_2 e^{-RT\Delta\psi/ZF}$, where ψ is the transmembrane potential [Eq. (2.5)]}. All cells, however, show very clear evidence that very many substrates that *can* cross the cell membrane are at a steady-state concentration inside the cell different from that outside. Sodium and calcium ions provide good examples, being at far lower concentration inside cells, in general, than outside [and this cannot be due to the Donnan potential (Section 3.7) which is in a direction such as to concentrate cations within the cell]. Amino acids are often at far higher concentrations inside cells than outside and yet cross cell membranes readily (by specific transport systems). Sugars, too, are concentrated within certain cells. Clearly, in all these cases, a type of transport system other than those that we have been considering must be involved. We term such concentrative systems, which drive substrates against their electrochemical gradient, *active transport* systems.

There are two main classes of these substrate-concentrating systems. In one class, the *primary active transport* systems, substrates are pumped across cell membranes by a process directly linked to the consumption of metabolic energy. Generally this involves the splitting of ATP, but it can also involve absorption of light energy or else harnessing the flow of electrons that takes place during metabolic oxidation. In the second class, *secondary active transport*, the flow of the substrate that is being pumped *up* its electrochemical gradient is coupled to the flow of a second substrate *down* its electrochemical gradient; the gradient of this second substrate is maintained by a primary active transport system. In the living cell a vast array of primary and secondary transport systems are operating, across the outer plasma membrane and across the membranes within the cell that bound the intracellular organelles, producing a multitude of electrochemical gradients for a wide variety of substrates. In this chapter we shall explore the properties of the secondary transport systems, as these are simpler to understand. In the next chapter we will study the primary active transporters.

5.1. COUNTERTRANSPORT ON THE SIMPLE CARRIER

In certain, defined circumstances even the simple carrier of Fig. 4.6 can establish a (temporary) concentration gradient. Consider Fig. 5.1, which shows an experiment performed using the glucose transporter that we discussed in Chapter 4. Remember that this transporter carries glucose *or* galactose; both are good substrates for it. In the experiment depicted, a sample of red blood cells was preequilibrated with 85 mM glucose. At time zero in Fig. 5.1, the cells were suspended in galactose at a concentration of 4 mM and the entry (and then exit) of galactose from the cells followed as a function of time. (The movement of galactose is plotted as the hatched bars on the histogram.) Note that galactose enters the cells, very soon reaching concentrations far higher than the concentration outside the cell, and is *concentrated* within the cells. The solid bars depict data obtained at the same time and show the loss from the cells of the glucose that was originally present. The entry of galactose is *coupled* here to the exit of glucose.

Why is the galactose concentrated within the cells, and what brings about the coupling between influx and efflux of the two sugars? The K_m for glucose transport on the red cell system is about 10 mM in the conditions of Fig. 5.1. Thus at the inner face of the membrane the system is fully saturated with the glucose. The galactose that enters the cell cannot escape, since it cannot compete with the vast excess of glucose. Yet, at the extracellular face, only galactose is present and, therefore, this sugar *can* enter the cell from that face. As the system is a carrier, it has the binding sites for sugar at one face of the membrane *shielded* from those at

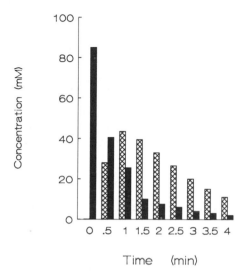

Fig. 5.1.   Countertransport of two sugars across the red cell membrane. Human red blood cells in saline at 20°C were originally preloaded with glucose (85 m*M*) and then transferred to a medium containing 4 m*M* galactose and no glucose. The glucose still left within the cell (solid bars) and the galactose that had entered (hatched) were measured. These are plotted as a histogram. By one minute the galactose had been concentrated within the cell over 10-fold. It left the cell more slowly than did glucose since the measurements here are of a net flux; glucose efflux is essentially one-way, while galactose efflux is a balance between the continued influx and the efflux. Based on data of G. F. Baker and W. F. Widdas (1973). *J. Physiol. (London)* **231**, 143–165.

the other face. Thus galactose can enter in spite of an outward flow of glucose. It is competition that ensures a one-way transport of the galactose. The flow of this sugar is coupled to that of glucose because both ride on the same carrier, and the reorientation of carrier conformations is brought about as each sugar moves down its concentration gradient.

This phenomenon—the coupling of the flow of one substrate to the flow of another moving in the opposite direction—is known as *countertransport*. Widdas predicted that carriers should behave in this way, and the phenomenon was later demonstrated experimentally by Walther Wilbrandt and Thomas Rosenberg. All carriers will behave like this. All that is needed is that the carrier be saturated at one face with one substrate, this substrate being absent at the opposite face where the second substrate, that to be concentrated, is situated. Not only is the phenomenon predicted by the carrier model. Its existence for any substrate–transporter system is good evidence that that system behaves according to the carrier model of Fig. 4.6, with its reorientation of the conformations of the carrier, under the influence of the substrate.

Concentration of substrate by countertransport on the simple carrier is only a temporary affair. There is, during the experiment, a net flow of substrate out of the cell on the carrier. Consider Fig. 5.1 again. By 3 min, much of the glucose that was in the cell has left it and the sugar-binding site at the cell interior is no longer saturated. Competition between galactose and glucose has decreased so as to be almost insignificant, and the ability of the system to maintain a concentration gradient is quickly disappearing.

5.2. EXCHANGE-ONLY SYSTEMS: THE ANTIPORTERS

In countertransport on the simple carrier, net transport will eventually bring about the disappearance of the concentration gradient of the driving sugar and hence the cessation of countertransport. This occurs because the unloaded carrier can reorient and bring the substrate-binding sites back to the high-concentration side, for a further cycle of net transport. In many carriers, however, the unloaded carrier cannot reorient at all across the membrane. There can be no *net* transport of a single substrate that uses such a system; hence these are *exchange-only* systems. All substrate transport on such an exchange-only carrier is strictly coupled, and the gradient for one substrate can diminish only if that for another substrate is built up. The formal model for such an exchange-only system, termed an *antiport system* or *antiporter*, is depicted in Fig. 5.2(a). An experiment showing the existance of a Na^+/Ca^{2+} antiporter in membrane vesicles isolated from muscle cells is seen in Fig. 5.2(b). With a sodium gradient present (i.e., with *potassium* inside the vesicles), the membrane vesicles accumulate calcium. When the gradient is reduced (i.e., when sodium is present both inside and outside the vesicles), calcium escapes from them.

5.2.1. The Kinetics of Antiport

Either substrate, S or P, can cross the membrane on an antiporter, but only in exchange for a second substrate molecule, either S for S (self—or homoexchange) or S for P (heteroexchange).

Substrate S moves from side 1 on the antiporter at a rate proportional to its concentration S_1; P moves from side 2 at a rate proportional to P_2. Thus, an exchange of S, coming from side 1, with P, coming from side 2, will occur at a rate proportional to the product of S_1 and P_2. In similar fashion, the exchange of S at side 2 with P at side 1 occurs at a rate

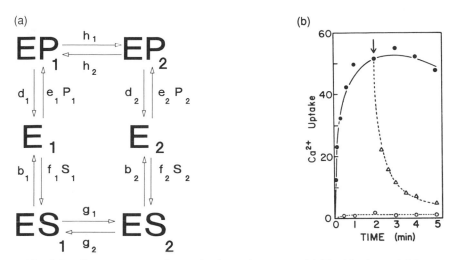

(a)   (b)

Fig. 5.2. Transport on exchange (antiporter) systems. (a) The kinetic model for an antiporter: E, the antiporter, S and P, its two substrates; ES and EP, the complexes between transporter and the substrates. Subscripts 1 and 2, respectively refer to sides 1 and 2 of the membrane. The rate constants for the various conformation changes and binding reactions are depicted as the symbols b, d, e, f, g, h with the appropriate subscripts. (b) Sodium and calcium exchanging on the sodium–calcium antiporter of heart muscle cells. At zero time, the sodium-loaded membrane vesicles were diluted into a medium containing either 160 mM KCl (solid circles) or 160 mM NaCl (open circles). At the arrow, 40 mM NaCl was added to the external medium. The units of Ca^{2+} uptake are nmol/mg (nanomoles per milligram) of protein. Taken, with kind permission, from J. P. Reeves *et al.* (1980) *in* "Calcium-Binding Proteins" (F. L. Siegel *et al.*, eds). Amsterdam, North Holland.

proportional to $S_2 \times P_1$. Exchange of S with S or P with P has no effect on the *net* transport of S and P. Thus, at the steady state, where the net transport of S from side 1 to side 2 is equal to that from side 2 to side 1 we have, for uncharged substrate, that

$$S_1 P_2 = S_2 P_1 \tag{5.1}$$

or,

$$S_1/S_2 = P_1/P_2 \tag{5.2}$$

Equation (5.2) is the *fundamental rule of antiport:* at the steady state, the concentration ratio of the one substrate that uses the antiporter is exactly equal to that of the other. Take the case where sodium and calcium both use the same antiporter [as is the case for the muscle cell membranes depicted in Fig. 5.2(b)]. Using the primary transport system for sodium (which we will discuss in Chapter 6), the concentration of sodium [S in Eq. (5.2)] is held about 10-fold lower on the inside of the cell (side 1) than on the outside (side 2). As a result of the presence of the antiporter, the concentration of calcium [P in Eq. (5.2)] will similarly be lower at side 1 than at side 2. Indeed, as we shall see in Section 5.2.4, the special properties of the antiporter of the muscle enable the intracellular calcium to be kept at a concentration gradient of several thousandfold, inside low. Being able to control the low level of cytoplasmic Ca^{2+} concentration is an important aspect of muscle function. Initiation of contraction of the muscle and its subsequent relaxation occur as a result of changes in the ambient calcium concentration, brought about by calcium channels and pumps.

Just as the simple carrier systems of Chapter 4 could be *electrogenic* (recall Section 4.7), so, of course, can the antiporters. Indeed, in the case just considered where sodium is exchanged for calcium ions, a complete cycle of net transport on this antiporter carries three sodium ions across the muscle cell membrane in one direction for every (doubly charged) calcium ion carried in the other. There is a net transfer of one positive charge in a cycle. There is a similar situation in the important antiporter that exchanges ADP and ATP (adenosine diphosphate and adenosine triphosphate) across the membrane of the mitochondrion. Here, the ADP bears three negative charges at neutral pH; the ATP, four such charges. A cycle of exchange of ADP with ATP transfers one unit of charge across the membrane, and the transfer rates and steady-state concentration ratios will be determined by the transmembrane potential of the mitochondrial membrane. The fact that this is inside negative ensures that at the steady state ATP is concentrated outside the mitochondrion more than is ADP, i.e., that the mitochondrion exports ATP in exchange for ADP, just as is required for effective functioning of the cell.

To correctly account for this transfer of charge by an antiporter, we must modify Eq. (5.2), to include the transmembrane potential $\Delta\psi$ and the charge Z_s on substrate S and charge Z_p on substrate P (see Box 5.1 for proof and Box 5.3 for the application of the derived equation to the Na^+/Ca^{2+} antiporters just discussed), obtaining

$$\ln(S_2/S_1) = \ln(P_2/P_1) + (Z_s - Z_p)F\,\Delta\psi/RT \qquad (5.3a)$$

or

$$2.303 \log(S_2/S_1) = 2.303 \log(P_2/P_1) + (Z_s - Z_p)F\,\Delta\psi/RT \qquad (5.3b)$$

Note that Eq. (5.3a or b) reduces to Eq. (5.2) when the net charge transferred across the membrane in a cycle ($Z_s - Z_p$) is zero, since the antiporter is in that case not electrogenic.

Box 5.1. Electrogenicity and the Antiporters

The proof of Eq. (5.3a or b) follows directly from Eq. (2.3), which we will rewrite in a modified form below. Recall that the chemical potential of a substance S is a measure of its capacity to perform work. We denote $U_{s,i}$ as the chemical potential of the solute S (where the subscript i is the side of the membrane being referred to, 1 or 2), U_s° as the standard-state chemical potential [i.e., the chemical potential of S at a concentration of 1 molal (i.e., 1 mol of S per 1000 g of water) at 0°C, and at zero electrical potential], V_s as its partial molal volume (i.e., the increment in volume per mole of solute added when an infinitesimal amount of S is added to the solution), $\mathbf{P}$ as the pressure exerted on the solution (in excess of atmospheric pressure), ψ_i as the electrical potential at which S finds itself at side i and, finally, Z_s as the valence of the ion S. We then can write the expression for S at side i

$$U_{s,i} = U_s^\circ + RT \ln S_i + V_s\mathbf{P} + Z_sF\,\psi_i$$

To obtain the corresponding equation for the second substrate P, we write P for S throughout in the preceding equation. We now calculate the change in chemical potential on moving n_s moles of substrate S from side 1 of the membrane to side 2, and n_p of substrate P in the opposite direction. The standard-state chemical potentials are, of course, unchanged during such a transfer as are, we assume, the pressure terms. At the steady state, the overall change in chemical potential for the coupled transport of S and P is zero. Summing the two equations formed by substituting appropriately in the modified Eq. (2.3) and equating the sum to zero, we obtain

$$n_sRT \ln(S_2/S_1) + n_sZ_sF\,\Delta\psi_{2\to1} + n_pRT \ln(P_1/P_2) + n_pZ_pF\,\Delta\psi_{1\to2} = 0$$

which simplifies to

$$\ln(S_2/S_1)^{n_s} = \ln(P_2/P_1)^{n_p} - (n_pZ_p - n_sZ_s)F\,\Delta\psi_{1\to2}/RT$$

In the case where the number of moles of S and of P moved are the same, $n_p = n_s$, and we recover Eq. (5.3a).

Consider the calcium–proton antiporter of the bacterium *Escherichia coli*. This exchanges one proton for one (doubly charged) calcium and is, therefore, electrogenic. A single charge is transferred across the membrane for each transport cycle. As we saw in Section 2.2, at 25°C a transmembrane potential of 59 mV will give a transmembrane concentration ratio of 10-fold for a univalent ion. According to Eq. (5.3a or b), the concentration ratio of calcium (outside/inside) will be greater than that of proton by just this same factor of 10, if it is exchanged for the singly charged proton in the calcium–proton antiport cycle. [The second term on the right-hand side of Eq. (5.3b) is equal to 1, the antilog of 10.] This is one example of the many proton antiporters that occur in bacteria. The full transport equation for an antiport system [corresponding to that given for the uniporters as Eq. (4.3)] is given in Stein (1986, pp. 308, 309).

5.2.2. Slippage and Leakage in Coupled Transport Systems

Equations (5.2) and (5.3a or b) set the maximum concentration ratio of P that its antiport with S can bring about (at any given ratio of S_2 to S_1 and $\Delta\psi$). If there is no other pathway for the movement of P, through either the antiporter itself (i.e., were the system to allow *some* net movement of P) or a parallel system, Eqs. (5.2) and (5.3a or b) will define the concentration ratio of P exactly.

It is helpful, in this context, to introduce two new terms into our discussion. The first is *leakage*, which we can define as any movement of a transported substrate via a pathway other than the one being immediately considered. Thus if two substrates are transported by an antiporter, movement of either of them on any other, parallel system is termed *leakage*. This could involve movement by simple diffusion, by some other simple carrier or some other antiporter. The second term to be introduced is *slippage*. This is movement of the transported substrate on the transporter that we are considering, uncoupled to the flow of the second substrate. If, for instance, in the antiporter model of Fig. 5.2, there is a small rate of reorientation of the unloaded carriers (banned on the model as drawn, but liable to occur in a real system!), net flows of both S and P will occur. The system will show slippage, and the concentration gradients of the two coupled substrates will be dissipated.

Equations (5.2) and (5.3a or b) describe the steady-state concentration ratios brought about by antiport coupling in the absence of leakage and slippage. Any real system will show some measure of each of these two phenomena. Thus Eqs. (5.2) and (5.3a or b) describe an idealized state, rather than the actual behavior of any real system. In many natural systems, evolution has selected for a high degree of coupling, and hence for a low degree of slippage. In some systems (we shall discuss some of these in

Chapter 7) leakage is controlled and the intracellular concentration of particular transported substrates can be regulated in this manner. Leakage is minimized by selecting for highly impermeable membranes in those systems for which efficient antiporters are crucial for cell function. The mitochondrion, for example, possesses many antiport systems that are coupled to proton flows. Appropriately, the basal proton permeability of the mitochondrial membrane, in the absence of antiporters, is one of the lowest known.

5.2.3. Asymmetry of Antiporters

Just as the simple carrier of Fig. 4.6 could behave asymmetrically, so can the antiporters of Fig. 5.2 be asymmetric and, indeed, many of them are. (The system that exchanges ATP and ADP across the membrane of the mitochondrion is just one example of an asymmetric antiporter.) This asymmetry is seen in the half-saturation parameters K_m and not in the V_{max} term.

The asymmetry (measured as the ratio of the K_m values) is given exactly by the *free-energy* difference between the two conformations of the free antiporter E_1 and E_2. (We obtained a similar result in Section 4.6.2 for the carrier.) Since this free-energy difference is the same whatever substrate uses the antiporter, all substrates that share the same antiporter are asymmetrically transported by this antiporter across the membrane to exactly the same extent, kinetically.

5.2.4. How the Stoichiometry of Substrate Binding Determines the "Intensity" of Concentration

Some antiporters bind more than one molecule of a particular substrate. Thus a sodium–calcium antiporter found in the plasma membrane of many cells moves three sodium ions in one direction, one calcium ion in the other. The *stoichiometry* of ion binding here is $Na:Ca = 3:1$. How does this affect the steady-state levels of Na and Ca? To find out, return to the discussion around Eqs. (5.2) and (5.3a or b). There we assumed that the stoichiometry of transport was $1:1$. However, as in Box 5.1, we can make a more general hypothesis, taking the stoichiometry of binding of S and P to the antiporter as s and p moles, respectively. Then the exchange of S and P from side 1 to side 2 will be proportional to $S_1^s \times P_2^p$, while that in the $2 \rightarrow 1$ direction will be proportional to $S_2^s \times P_1^p$. At the steady state these rates will be equal, so we obtain a relation analogous to Eq. (5.2) such that, at the steady state

$$S_1^s/S_2^s = P_1^p \times P_2^p$$

or,

$$(S_1/S_2)^s = (P_1/P_2)^p \tag{5.4}$$

Thus, what one might term the *intensity* of concentration (how one concentration ratio depends on the power of another) depends on the stoichiometric terms, s and p. For the sodium–calcium antiporter of muscle, mentioned in Section 5.2.1, take S to be sodium with $s = 3$, and P to be calcium with $p = 1$, and substitute in Eq. (5.4). Then a 10-fold concentration gradient of Na will elicit, at the steady state, a $10^3 = 1000$-fold concentration gradient of Ca. This seems like getting energy out of nothing, but one has, of course, to move *three* sodium ions for each calcium ion during the overall cycle of calcium ion pumping. The situation is rather like an electrical transformer in which a low voltage can be transformed into a high voltage, but only at the expense of a high current being consumed to produce a low transformed current. The product (voltage $\times$ current) is constant, as this is a measure of the electrical energy produced or consumed.

Box 5.1 gives a full derivation of the antiporter equilibria, which includes also the terms involving the transmembrane potential, $\Delta\psi$, and the net charge carried by a complete cycle of antiport.

5.2.5. Some Particular Antiporter Systems

Many cells contain antiport systems. Most common are proton–sodium antiporters, calcium–sodium antiporters, and the chloride–bicarbonate antiporter. This latter system, which does indeed carry out a slow net transport of its substrates, we discussed as an example of a facilitated diffusion system in Section 4.8.1. Below we discuss the two sodium-dependent systems in greater detail.

5.2.5.1. THE Na$^+$/H$^+$ ANTIPORTER AS A TRANSDUCER
         OF CELL-TO-CELL SIGNALS

Most animal cells have an internal pH that is rather close to the extracellular pH. This is a surprising finding. After all, most cell membranes are relatively permeable to protons, so that one would expect the Donnan potential (Section 3.7) to govern the proton distribution. Protons might be expected to distribute like potassium ions (10-fold higher within the cell than outside it), giving a pH about one unit lower inside the cell than out.

However, a sodium–proton antiporter couples an inward flow of sodium ions to an outward flow of protons, so that the level of protons at the steady state is lower than the Donnan distribution would predict. On the other hand, the level is not that to be expected were the antiporter rule of Eq. (5.2) to be solely responsible for the proton gradient. On its own, the Na^+/H^+ antiporter would couple protons and sodium so that the proton concentration would be some 10-fold *lower* inside the cell than outside, just as is the case for sodium ions. Clearly, the pumping capacity of the antiporter is offset by a leakage pathway for protons. Protons move into the cell by this leakage route as fast as the Na^+/H^+ antiporter pumps them out. The level of internal pH is a balance between the rate of pumping out of the protons on the antiporter and the rate of entry by leakage. The internal pH is thus set by the relative values of the rate constants of proton pumping and proton leakage.

The Na^+/H^+ antiporter is inhibited by micromolar concentrations of amiloride (a drug used to induce diuresis, i.e., enhanced urine production by the kidneys), a fact that has been much made use of in the study of the system. Sodium fluxes inhibitable at such concentrations of amiloride are likely to be carried on the Na^+/H^+ antiporter. (At lower concentrations $(10^{-7} M)$, amiloride inhibits sodium movement on sodium channels, so the criterion of amiloride inhibition must be used with care.)

When cells are stimulated to divide by the addition of *growth factors* (such as certain hormones; see any cell biology textbook), one of the first events to take place is a *rise* in the intracellular pH of approximately 0.2–0.3 pH units [which can be measured by pH-sensitive fluorescent dyes; see Fig. 5.3(a)]. This rise in pH is interpreted by a range of intracellular enzymes as a signal to change their activity. Cells that undergo such a rise in internal pH enter into a new state, that of a cell preparing for division, and demonstrate a manifold change in behavior. From the discussion we have just had above, we can see that a rise in intracellular pH would be brought about if (1) the Na^+/H^+ antiporter were to be activated to operate more rapidly or (2) the leakage pathways of the cell membrane for protons were to be blocked. The rise in pH is blocked by moderate concentrations of amiloride, which we have seen to be a fairly specific inhibitor for the Na^+/H^+ antiporter. Also, the increase in internal pH does not occur if sodium is omitted from the extracellular medium. Thus it seems that it is, indeed, activation of the Na^+/H^+ antiporter that brings about the rise in pH. Calcium entry also activates the antiporter [see Fig. 5.3(a)].

When the growth factor binds to specific receptors in the cell membrane, a train of events occurs, one of which results in the change of conformation of the Na^+/H^+ antiporter and an increase in its affinity for protons at the intracellular face [see Fig. 5.3(b)]. It seems that there is a

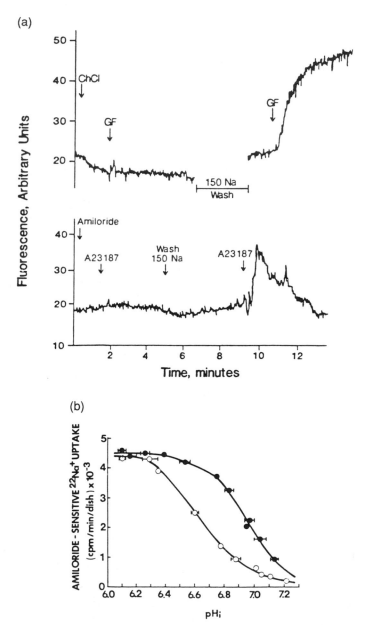

Fig. 5.3.   Stimulation of antiporters by growth factors. (a) Human fibroblast cells were grown on glass coverslips and then deprived of growth factors and loaded with the pH-sensitive fluorescent probe dimethylcarboxyfluorescein. An upward deflection of the fluorescent signal indicates alkalinization of the cellular cytoplasm. At the times marked GF (upper curves) they were treated with epidermal growth factor (EGF) in the absence (left)

special proton-binding site at the intracellular face of the antiporter that regulates the activity of the system. With the site blocked by a proton, the antiporter is inhibited. In resting cells, the antiporter is largely (but not totally) inhibited at the level of pH prevailing in the cell. However, stimulation of the antiporter by growth factors shifts the pH–activity curve to the right [Fig. 5.3(b)] so that the system is now fully active at the pH prevailing within the cell, although this pH has itself not yet changed. All the transporters that were held empty and idle at the intracellular face of the membrane are now able to bind protons, transporting these out of the cell. Transport continues until a new balance is established between this enhanced rate of outward pumping and the inward leak. The activation of the antiporter is a transient affair and the rise in intracellular pH soon reverses, although the result of its action in bringing the cell into a new "set" lasts for a sufficient time to ensure the completion of a round of cell division. We will see in Chapter 7 that this Na^+/H^+ antiporter is also involved in the ion movements that occur during regulation of the volume of cells and in transport across epithelia.

5.2.5.2. ROLE OF THE Na^+/Ca^{2+} ANTIPORTER IN THE REGULATION OF INTRACELLULAR CALCIUM

Like that of protons, the concentration of intracellular calcium is not that given by the Donnan potential. Rather than intracellular Ca^{2+} being higher than extracellular, it is lower, indeed about 10^4-fold lower in many cells. (Intracellular and extracellular Ca^{2+} levels for cardiac muscle cells are recorded in Box 5.2.)

Two systems, working in parallel, bring about this low intracellular Ca^{2+} concentration (Fig. 5.4). One system is an ATP-splitting primary transporter of Ca^{2+} that we shall discuss in Section 6.3. The other is a Na^+/Ca^{2+} antiporter that we have mentioned a number of times [see Fig. 5.2(b)]. In muscle cells, this system has a stoichiometry of 3 Na^+ to 1 Ca^{2+}. With the concentration gradient for sodium being about 10-fold (inside low), such a stoichiometry would yield a 1000-fold ratio for

---

and then presence (right) of sodium ions. In the lower curves, cells were treated with the calcium ionophore A23187 at the indicated times, in the presence (left) and absence (right) of amiloride, an inhibitor of the Na^+/H^+ antiporter. Taken, with kind permission, from L. L. Muldoon, R. J. Dinerstein, and M. L. Villereal, (1985). *Am. J. Physiol.* **249**, C140–C148. (b) Effect of growth factors on the pH dependence of the Na^+/H^+ antiporter. The growth of Chinese hamster lung fibroblasts was arrested by serum deprivation and then growth reinitiated by the addition of growth factors. The pH–activity curve for amiloride-inhibitable Na^+ uptake was determined with (solid circles) or without (open circles) these growth factors. Taken, with kind permission, from S. Paris and J. Pouyssegur, (1984). *J. Biol. Chem.* **259**, 10989–10994.

Box 5.2. Levels of Calcium Ion in Cardiac Muscle Cells

Measurements were made on cultured embryonic chicken heart cells [data from Lieberman *et al.* (1984)]. The levels of ions found are given in the following tabulation.

Measurement	K^+	Na^+	Cl^-	Ca^{2+}
Extracellular medium (mM)	5.4	144	128	2.7
Intracellular fluid (mM)	133	16	25	2.3×10^{-4}
Equilibrium potential (mV)	-83	$+57$	-42	$+122$

The equilibrium potential is calculated using Eq. (2.6). The measured membrane potential under these conditions is -80 mV. This means (Section 2.2) that only the potassium ion is at equilibrium with the transmembrane potential. The sodium ion level is kept low by the sodium pump (see Section 6.1) and the calcium level kept low by the combined action of the sodium–calcium antiporter and the calcium pump (see text, Section 5.2.5.2). Movement of chloride ions is also in part linked to an active system whose nature is not clearly understood as yet.

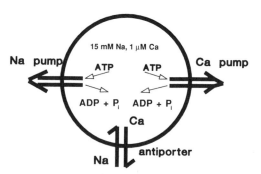

Fig. 5.4. Interactions of sodium and calcium pumps and antiporters. This conceptual scheme shows two ATP-dependent pumps (see Chapter 6), one for sodium and one for calcium, which pump these ions *out* of the cell. In addition, a sodium–calcium antiporter exchanges (three) sodium ions for one calcium ion. The final electrochemical gradient of calcium is thus determined by the combined action of a high-affinity, low-capacity ATP-dependent pump and a low-affinity, high-capacity antiporter.

calcium, again inside low. We find instead a 10,000-fold ratio. Clearly an additional factor must be operating. This is, of course, the transmembrane potential, which, in the muscle cells, is approximately 80 mV, inside negative. As we saw in Section 5.2.1, the Na^+/Ca^{2+} antiporter is electrogenic, transferring a positive charge inward in a complete cycle in which three sodium ions enter the cell in exchange for a (divalent) calcium ion that is extruded. The combined effect of the concentration ratio of the sodium and the transmembrane potential would be sufficient to account for the calcium ion concentration ratio (Box 5.2). When the sodium pump is blocked, the calcium concentration rises in accordance with the predictions of the antiporter model.

What is the advantage to the cell of having both the Na^+/Ca^{2+} antiporter and the ATP-consuming calcium pump working in parallel? To understand this one must know that calcium ions, like protons, are of great importance in cell-to-cell signaling. Transmitter substances that pass between cells bring about the opening of specific calcium channels (such as those described in Chapter 3), allowing calcium ions to enter the target cell down their concentration gradient. This brings about a rise in intracellular calcium that, in turn, sets off a train of reactions within the target cell. When the need for the cell-to-cell signal has passed, the level of intracellular calcium ions must be restored to that low level characteristic of the resting state of the cell. This is the role of the two pumping systems. The problem, however, is that the rise in calcium ion concentration is often very large (see Box 3.9), with the consequence that a good deal of calcium must be pumped out to restore the resting-state levels. This pumping requires much energy. Pumping is performed by the two systems working in parallel (Fig. 5.4), and this leads to a lowering of the overall energy consumption. The Na^+/Ca^{2+} antiporter pumps out the great bulk of the calcium that has entered, setting the level of intracellular calcium to the concentration that it itself is capable of reaching, namely, a 10,000-fold gradient. The Na^+/Ca^{2+} antiporter has a high capacity for calcium transport but a comparatively low final effectiveness. The primary calcium pump, however, is a very effective pump, having a high stoichiometry of calcium ions pumped to ATP molecules consumed (see Section 6.3) and a high affinity for calcium at the cytoplasmic surface. The small number of such pumps present per cell cannot, however, cope with a large bulk of calcium ions. Their action is to scavenge out the last traces of calcium ions from within the cell, bringing the resting steady-state level down to tens of nanomolars. The overall effect is to pump out the great bulk of the calcium ions by a process involving a relatively low energy consumption per mole of calcium. The traces that remain are pumped out by a high-affinity, low-capacity system that must consume a good deal of energy for each mole of calcium that it transports.

5.3. THE COTRANSPORT SYSTEMS: TWO (OR MORE) SUBSTRATES THAT RIDE TOGETHER IN SYMPORT ON A SIMPLE CARRIER

One of the triumphs of modern physiology has been the working out of how sugars and amino acids are actively transported across the epithelia of intestine and kidney. We will consider, now, the research that led to the solution of this problem, and the far-reaching implications of this research for the understanding of active transport in many other systems: in animal, plant, and bacterial cells.

Free sugars and amino acids are found in the *lumen* of the intestine after the digestion of food. From the lumen, they are actively transported across the epithelia to the *serosal* surface, i.e., the surface that comes into contact with the blood supply. (The same substrates are present in the lumen of the kidney tubules following filtration of the blood in the initial stage of urine production. Again, they are absorbed from the lumen by active transport and returned to the blood.) One can show directly that movement of these substrates from lumen to blood is against the concentration gradient of sugar or amino acid. A simple technique for studying intestinal absorption is to use the "everted sac" preparation, introduced by Thomas Wilson. In this technique, a length of intestine (from a rat or hamster) is cut off, turned inside out (everted), filled with a solution of the substrate to be transported, tied at both ends [Fig. 5.5(a)] and then placed in an incubation bath containing the substrate, nutrients, and salts. The sac having been everted, what is now the outside was originally the surface facing the lumen in the intact animal, and it is from this surface that active transport occurs. Active transport takes place into the interior of the sac from which samples can be removed at intervals to study the process of absorption.

The time course of accumulation of an amino acid by such a preparation is depicted in Fig. 5.5(b). Here the concentration of the amino acid was measured over a 4-hr period (1) in the external bath [labeled "mucosal side" in the Fig. 5.5(b)], (2) within the sac, i.e., in the medium bathing the luminal face of the intestinal strip ("serosal side"), and, very revealingly, (3) *within* the tissue of the sac ("tissue"). One can see that the amino acid is transported into the tissue, reaching a level at least eightfold higher than that at the luminal face, and that from the tissue it moves to the sac interior. The active transport system must be located in the membrane that is between the lumen and the cell cytoplasm. Using such preparations, some crucial observations were made. First, it was shown that cellular metabolism was necessary for active transport to take place. If metabolism was blocked by the addition of inhibitors or by depriving the

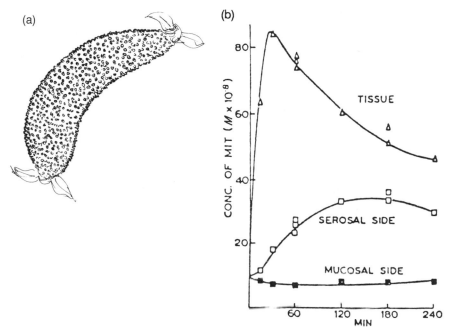

Fig. 5.5. Cotransport into intestinal sacs: the everted sac preparation. (a) An everted jejunal sac. Note the villi on the outside of the preparation. In life, these are on the inside of the jejunum. The sac has been tied with 1 ml of fluid inside it. (Original drawing by Chana Stein.) (b) A classical experiment: changes in concentration of the amino acid analog monoiodotyrosine (MIT) *within* an everted sac (serosal side), in the *bath* (mucosal side), and within the intestinal *tissue* (tissue) during the time of incubation of a sac in 10^{-7} M of the amino acid. Taken, with kind permission, from D. Nathans, D. F. Tapley, and J. E. Ross, (1960). *Biochim. Biophys. Acta* **41,** 271–282.

tissue of necessary metabolites, no active transport could be seen. Second, it was found that the presence of *sodium* ions was absolutely essential for active transport. They could not be replaced by lithium, potassium, magnesium, or any other cation.

The role of sodium was clarified by some important observations made by the physiologist Robert Crane and co-workers. They showed that a sac preparation such as that shown in Fig. 5.5(a) could be used to study the *exchange* of labeled glucose when unlabeled glucose had already come to equilibrium within the sac. This exchange of glucose was strictly dependent on the presence of sodium ions. Now, an exchange measurement does not measure active transport. The glucose has already been absorbed against its concentration gradient. One is here at the steady state and measuring simply the movement of glucose on its carrier into the cell. Crane's experiment showed unequivocally that *movement* of glucose via its carrier required sodium ions, and this fact led him to propose a general solution to the problem of how such active transport systems work. He suggested that the true substrate for the carrier at the luminal membrane is (glucose + sodium). If the carrier can transport glucose only *simultaneously* with sodium, then movements of the two substrates, sugar and sodium, will be strictly coupled on a *cotransport* or *symport* system. The *gradient* of glucose at the steady state will be dependent on the *gradient* of sodium ions.

We shall see in Chapter 6 that the gradient of sodium is established directly by a primary transport system, the sodium pump. It is clear, therefore, that in the living cell, the gradient of glucose is set by that of sodium and not vice versa.

5.3.1. Crane's Gradient Hypothesis

Crane's gradient or cotransport hypothesis is depicted in Fig. 5.6, a scheme developed in great detail by Georgio Semenza. Here S is the substrate, say glucose or an amino acid, and A is the ion (generally sodium in animal cells, or proton in bacteria) that is coupled to S for the active transport. Three forms of the model are shown: (a) where the order of binding of S and A is *random*, (b) where it is *ordered* and A binds first, and (c) ordered again, but where it is S that binds first. The free carrier E can reorient across the membrane, as can the doubly bound forms EAS, but singly bound forms ES and EA cannot reorient. It is easy to see that the rate of transport from side 1 to side 2 will be proportional to the concentrations of both A and S at side 1, i.e., to the product S_1A_1. Similarly, the rate of transport from side 2 to side 1 will be proportional to the product S_2A_2. At the steady state, when net transport ceases, these two rates will be equal so that

$$S_1A_1 = S_2A_2$$

or

$$S_1/S_2 = A_2/A_1 \tag{5.5}$$

Equation (5.5) is the *fundamental rule of cotransport* and shows that, at the steady state, the concentration ratios of the driving substrate and the driven substrate are inversely related to one another. If Na^+ is *low* within the cell, kept low by some primary active transport, a secondary cotrans-

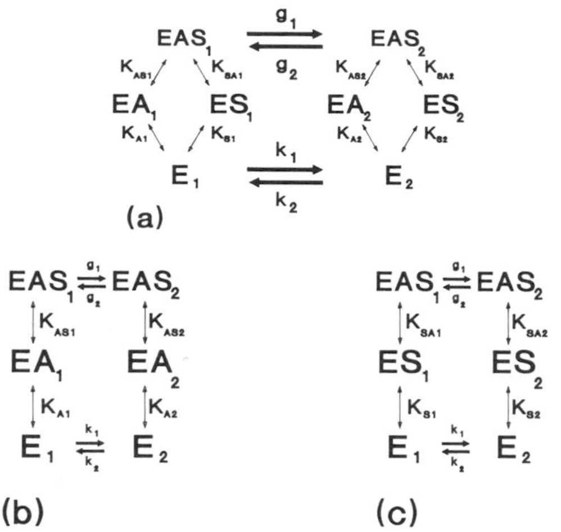

EAS$_1$ $\underset{g_2}{\overset{g_1}{\rightleftharpoons}}$ EAS$_2$

K_{AB1} / \ K_{BA1} K_{AB2} / \ K_{BA2}

EA$_1$ ES$_1$ EA$_2$ ES$_2$

K_{A1} \ / K_{B1} K_{A2} \ / K_{B2}

E$_1$ $\underset{k_2}{\overset{k_1}{\rightleftharpoons}}$ E$_2$

(a)

EAS$_1$ $\underset{g_2}{\overset{g_1}{\rightleftharpoons}}$ EAS$_2$

$\uparrow K_{AB1}$ $\uparrow K_{AB2}$

EA$_1$ EA$_2$

$\uparrow K_{A1}$ $\uparrow K_{A2}$

E$_1$ $\underset{k_2}{\overset{k_1}{\rightleftharpoons}}$ E$_2$

(b)

EAS$_1$ $\underset{g_2}{\overset{g_1}{\rightleftharpoons}}$ EAS$_2$

$\uparrow K_{BA1}$ $\uparrow K_{BA2}$

ES$_1$ ES$_2$

$\uparrow K_{B1}$ $\uparrow K_{B2}$

E$_1$ $\underset{k_2}{\overset{k_1}{\rightleftharpoons}}$ E$_2$

(c)

Fig. 5.6. Three kinetic schemes for cotransport: (a) the random model; (b) the ordered model with A binding first at both membrane faces; (c) the ordered model with S binding first at both membrane faces. Here E is the cotransporter and S and A, its two cosubstrates; EA and ES are the complexes with a single cosubstrate and cannot undergo the conformation change that allows the substrates to cross the membrane. The ternary complex EAS *can* undergo such a conformation change. Sides 1 and 2 of the membrane are symbolized by subscripts 1 and 2. The terms in g and k are the rate constants for the depicted conformation changes of the doubly loaded and free cotransporter, respectively. The terms in K are the dissociation constants for the equilibria between the cotransporter and its substrates as depicted.

port of sodium with glucose or an amino acid will keep that cotransported substrate at a *high* concentration within the cell.

Proof of Crane's cotransport hypothesis speedily followed. One can mount a flat sheet of intestine (a strip of intestine being cut open to form the sheet) between two compartments that can be loaded with different solutions and into which electrodes can be placed. This setup is known as an Ussing chamber after Hans Ussing (see Box 2.2), who first used it in studies of active transport across frog skin [Fig. 5.7(a)].

When glucose and sodium ions are placed in the solution bathing the luminal surface of such a preparation and the tissue is given nutrients and oxygen, not only can a flow of glucose across the epithelium be measured, but also an *electric current*, flowing in such a direction as to transfer positive charge from the luminal to the serosal surface. In Fig. 5.7(b) we see direct measurements of this electric current when a sheet of rabbit intestine formed the barrier across the Ussing chamber. The first arrow indicates when (the nonmetabolizable) sugar was added. An immediate increase in the current occurred, this reaching a steady state in less than 5 min. At the second arrow, a specific inhibitor of sugar transport was

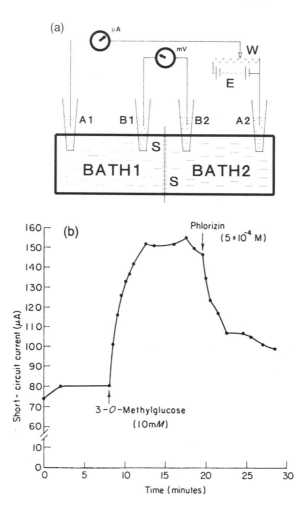

Fig. 5.7.   Ion movements across epithelial layers. (a) A schematic diagram of the classical Ussing chamber for measuring transepithelial ion flows. Baths 1 and 2 are separated by a barrier composed of a flat piece of epithelium (e.g., skin or intestine), SS in the figure. The electrodes B_1 and B_2 allow measurement of the transmembrane potential; electrodes A_1 and A_2 allow the potential from the Wheatstone bridge W and the power source E to be applied across the membrane, and the current flowing across it to be measured (W is often set so that the transmembrane potential is zero, the potential developed *by* the epithelium being just balanced by the applied voltage so that the current is that maintained by the ion pumps of the epithelium). Redrawn from Ussing (1954). (b) The current flowing across a piece of rabbit ileum, mounted as in (a) above. At the first arrow, a nonmetabolizable sugar is added. At the second arrow, an inhibitor of sugar transport. Addition of the sugar stimulates a flow of current that can be shown to be equal to the amount of sodium ions transported across the epithelium. Taken, with kind permission, from S. G. Schultz and R. Zalusky, (1964). *J. Gen. Physiol.* **47**, 1043–1059.

added (phloridzin), and a great reduction in sodium current occurred. The maximum flux of sodium ions in this preparation was some 3–5 μmol/cm^2 of surface per hour, while the maximum flux of sugar was 4–6 in these same units, suggesting a one-for-one movement of sugar and sodium. Thus glucose transport is stimulated by the presence of sodium, sodium transport is stimulated by the presence of sugar, and their transport is approximately in a one-for-one ratio, exactly the predictions of Crane's cotransport scheme of Fig. 5.6. (See Box 5.3 for an introduction to the molecular biology of the sodium–glucose cotransporter.)

Box 5.3. Molecular Biology of the Sodium–Glucose Symporter

It has been known for many years that the drug phloridzin is an effective inhibitor of sugar uptake across the epithelia of intestine and kidney. The drug blocks the sodium–glucose symporter in these tissues [Fig. 5.7(b)] and has been shown to bind tightly to cell membranes that contain the symporter. Interestingly, the binding of the drug to these membranes is itself sodium-dependent.

To study this binding further, monoclonal antibodies against pig kidney membranes were raised in rabbits. (Monoclonal antibodies are highly specific antibodies, produced by a single clone of antibody-producing cells; see any cell biology textbook). These antibodies can prevent the binding of phloridzin, presumably by blocking the active center of the symporter in these membranes. It was shown that these antibodies precipitated a 75-kDa polypeptide from solubilized membranes of these cells. Most importantly, these antibodies affected the binding to the 75-kDa protein of a chemical reagent, fluorescein isothiocyanate, and both sodium ions and glucose also stimulated this binding in a way that was itself affected by the binding of the antibody. Thus binding of the antibody seems to induce a conformational change in this protein. The 75-kDa protein appeared to be the sodium–glucose symporter.

Confirmation of this view has come from the cloning of the symporter. Ernest Wright and Mathias Hediger have developed a new cloning strategy. They used the oocytes of the frog *Xenopus* in which foreign messenger RNA molecules are expressed as proteins, some as membrane proteins (see Box 3.3). Suitably treated mRNA from rabbit intestine, injected into the frog oocytes, led to the expression of sodium-dependent, phloridzin-inhibitable glucose uptake in the oocytes. Wright and colleagues then *fractionated* the mRNA, obtaining a frac-

(Continued)

tion that was enriched in sequences that coded for the glucose symporter. From this fraction they prepared a cDNA clonal library and
screened this library for a clone that might contain the symporter gene.
They found a clone, whose RNA increased glucose uptake by the
oocyte more than 1000-fold. Sequencing the DNA of this clone led to
the identification of the cDNA for the gene, which coded for a polypeptide of molecular mass 73,080 Da, close to the mass of the protein
identified by antibody binding.

The hydropathy plot and the derived tentative structure of the transmembrane helices are depicted in Fig. 5.8. Most important is the find-

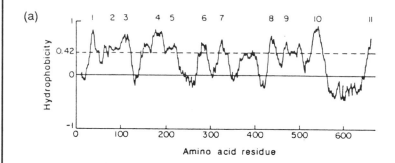

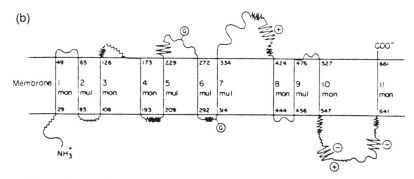

Fig. 5.8. Molecular structure of the sodium–glucose cotransporter. (a) Hydropathy
plot of the Na^+/glucose cotransporter as derived from the amino acid sequence deduced
by cloning of the gene. (b) Putative disposition of the transmembrane helices, deduced
from (a) and additional information. The components mon and mul are helices that
appear, from their amino acid sequences, to be present in monomeric or multimeric
regions; G represents a possible glycosylation site. The symbols + and − represent
clusters of charge. The wriggly lines represent different classes of secondary polypeptide
structures. The amino acid residues are numbered as are the transmembrane helices.
Taken, with kind permission, from Hediger *et al.* (1987).

(Continued)

ing that there is no homology between the sequence of the sodium–glucose symporter and that of the glucose uniporter discussed in Box 4.1. This is despite the fact that the glucose *uni*porter has homologies with several proton–sugar *sym*porters from bacterial cells.

Much subsequent work has fully validated Crane's elegant model for sugar and amino acid transport in intestine and kidney and for a wide variety of other substrates in other tissues. Apart from sodium, as we shall see, a second driving ion has been demonstrated in that protons are often the cotransported substrate, especially in plant and bacterial systems. For the proton systems, the term "symport" was historically used instead of "cotransport." The terms are exact synonyms, and here we use them interchangeably for both sodium- and proton-driven systems.

5.3.2. V and K Kinetics in Cotransport

When the current across the intestinal sheet shown in Fig. 5.7(b) is measured as a function of the glucose concentration at the luminal surface, at two different concentrations of sodium at that surface, the data in Fig. 5.9(a) are obtained. Sodium here affects the maximum velocity of sugar transport (V_{max}), whereas the half-saturation concentration (K_m) is unaffected. This is described as "V kinetics." Other types of behavior can be found. Figure 5.9(b) shows a clear finding of K kinetics in that here (for the uptake of sulfate ions cotransported with protons into a mold *Penicillium notatum*), K_m is greatly affected by the pH, while V_{max} is unchanged. Finally, in Fig. 5.9(c) (for calcium uptake as affected by pH in cotransport of calcium and proton in *Penicillium notatum*), both V_{max} and K_m are affected (V and K kinetics). All three types of behavior are to be expected from the cotransport model depicted in Fig. 5.6. Indeed, each phenomenon is a prediction of one of the different submodels of Fig. 5.6.

We can see this intuitively by understanding the line of argument presented in Sections 5.3.2.1 and 5.3.2.2. We assume in what follows that the transport steps—those in terms of g and k in the model—are slow in comparison with the rates of breakdown of the symporter–substrate complexes, the dissociation steps. Where this assumption has been tested, it has been found to be a reasonable one. Note that the maximum velocity of cotransport, V_{max}, on all models of Fig. 5.6 is then determined by the *concentration* of the form *EAS* multiplied by the *rate* of the conformation change that enables the substrate to cross the membrane. In contrast, K_m is determined by the probability that substrate and cotransporter will bind to each other, i.e., by the concentration of that form of the cotransporter that binds to the substrate in question.

(a)

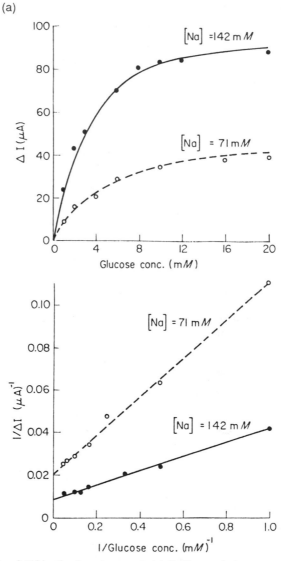

Fig. 5.9. *V* and *K* kinetics in cotransport. (a) Sodium and glucose cotransport across rabbit ileum, measured using the setup depicted in Fig. 5.7. The dependence on the glucose concentration of the current flowing (a measure of the sodium flux) was determined at two different concentrations of sodium. The bottom half of the figure is a reciprocal plot (see Section 4.2) of the data in the upper half. Taken, with kind permission, from S. G. Schultz and R. Zalusky, (1964). *J. Gen. Physiol.* **47**, 1043–1059. The data in (b) and (c) show kinetics of sulfate uptake by the mold *Penicillium notatum* at room temperature. The data in (b) show the mutual effects of sulfate and protons at a fixed level (20 mM) of calcium ions, those in (c) protons and calcium ions at a fixed level (1 μM) of sulfate. Reprinted with permission from J. Cuppoletti and I. H. Segel, (1975). *Biochemistry* **14**, 4712–4718. Copyright, 1975, American Chemical Society.

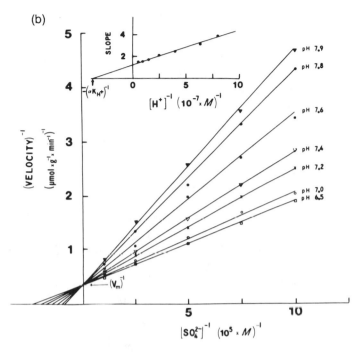

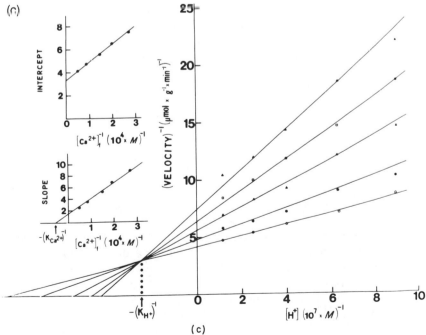

Fig. 5.9.   (*Continued*)

5.3.2.1 *K* KINETICS

Consider first the model where it is A that binds first to the cotransporter [Fig. 5.6(b)], while it is S whose concentration is continuously varied. Now at any concentration of A, a certain fraction of the free carrier will have been complexed in the form EA and be available for binding to S and the subsequent transport event. If only a low concentration of S is present, only a small amount of the complex EA will form EAS and be transported. If, however, S is available in high concentrations, any EA formed is immediately complexed as EAS, and both A and S are transported. More free E is converted into EA to replace that which has been removed as EAS. With even a small fraction of total carrier present as EA at any time, a sufficiently high concentration of S will be able to bring *all* the cotransporter into the form EAS. Since all the cotransporter is now in the form EAS, and engaged in transporting the substrates across the membrane, there can be no experimental situation that will give a higher velocity of transport. Thus V_{max} will be independent of the concentration of A. But the amount of S required to half-saturate will be totally dependent on A. If only a little A is present, only a small amount of EA will be available at any time, and the chances of encountering EA will be small. Therefore a high concentration of S will be required to drive the equilibrium $E + A = EA$ into the form EA for combining with S to give EAS. Thus K_m for S will be large. If a great deal of A is available, much EA is present and only a little S will be needed to drive the equilibrium to EAS. Thus K_m is small. The varying K_m of the K kinetics is an indication that the substrate whose concentration is varied continuously is the second to be bound.

5.3.2.2. *V* AND *K* KINETICS

Now consider the complementary case where it is again S whose concentration is continuously varied, but S itself is the first substrate to be bound [Fig. 5.6(c)]. Then the maximum velocity of transport will be totally dependent on the concentration of A. However much S is present—even if all E is converted into ES—only that amount of ES as can be converted into EAS according to the prevailing concentration of A will contribute to transport of S (and A). The V_{max} value will be strictly dependent on A. As before, the K_m for S will decrease with the concentration of A, because a high concentration of A drives the equilibrium between E and S toward formation of ES by removing the product of this reaction, ES. (One needs to think this through clearly. Since A removes ES, a high concentration of A actually reduces the level of ES, while yet driving the formation of ES. At any given time, however, the concentration of ES is lower if A is higher). Hence we find V and K kinetics.

The astute student will have realised that, for any system, if K kinetics is found for one of the two cosubstrates, V and K kinetics must be found for the other, since if one is bound first to the carrier, the other *must* be bound second.

However, K kinetics cannot be obtained for the random binding model of Fig. 5.6(a), so the finding that K kinetics applies is an indication that binding of the substrates is ordered. In one particular case of the random binding model, V kinetics is characteristically found.

5.3.3. Cis and Trans Inhibition Between Cosubstrates as Tests of the Cotransport (Symport) Model

Since the true substrate in cotransport is the substrate pair $S + A$, it is a simple matter to write down the kinetic equations that describe cotransport. One merely rewrites the fundamental carrier model equation [Eq. (4.3)] substituting $A \times S$ for S everywhere. (Transport depends on the joint probability of A and S being bound to the symporter. Therefore, transport depends on the *product* of A and S terms.) The result obtained is recorded in Table 5.1, in terms of resistances in R and half-saturation constants in K. The interpretation of these experimentally derivable parameters in terms of the model of Fig. 5.6 is listed in Table 5.1. Just as for the simple carrier of Fig. 4.6, one can devise for the cotransporter experimental procedures for the zero trans, infinite trans, and equilibrium exchange experiments. Some of these experiments can have quite bizarre outcomes, of which the student should be aware.

Consider, for example, an experiment in which the order of addition or breakdown of the EAS complex at side 2 is that depicted in Fig. 5.6(b), i.e., where it is S that leaves the EAS complex first, or binds last to E. Then if cells are loaded with labeled A at side 1, and the rate of exchange of A with unlabeled A at side 2 is studied, this rate will depend on S_2, the concentration of S at side 2. With no S at side 2, EAS_2 will break down to deliver S and then EA_2 will break down to deliver A. If, however, the concentration of S is high at side 2, little EA_2 will be present to break down. Any EA_2 formed will immediately react with S_2 and re-form EAS_2, which will return to side 1, *without delivering any A at side 2*. Thus the rate of transport of labeled A from 1 to 2 will decrease if the concentration of S is increased at side 2. We have *trans inhibition*, an unusual phenomenon, predicted by the cotransport model and indeed often found experimentally. (An example of trans inhibition, for a bacterial system, is shown in Fig. 5.13, below.) Similarly, the complementary phenomenon is predicted and has been found: in the experiment performed in the reverse direction, uptake of labeled A *from* side 2 will diminish as the concentration of S is increased at side 2. This is *cis inhibition*. It seems at first sight a contradiction that a cotransport system should show cis inhibition between its two substrates, but consideration of Fig. 5.6(b) shows that this will indeed be the situation where it is exchange that is being measured, i.e., where the carrier that is reorienting to side 2 is doing so in the form EAS rather than in the form of free carrier, E. In contrast, uptake of S, the second substrate to bind at side 2, will be stimulated by the presence of A at side 2, just as in a zero trans cotransport. This is because the formation of EAS at side 2 is dependent on the prior formation of EA_2, which will depend on how much A is present. One will find cis inhibition *and* cis stimulation between the two cosubstrates in an exchange situation.

TABLE 5.1

Kinetic Solutions for Various Forms of the Simple Cotransporter[a,b]

$$v_{1\to2}^S = \frac{KS_1A_1 + S_1A_1S_2A_2}{K^2R_{00} + KR_{12}S_1A_1 + KR_{21}S_2A_2 + R_{ee}S_1A_1S_2A_2}$$

	Model of		
	Fig. 5.6(b), A binding first	Fig. 5.6(c), S binding first	Fig. 5.6(a), random order of binding
$nR_{12} =$	$(1/k_2)(1 + A_2/K_{A2}) + 1/g_1$	$1/k_2 + (1/g_1)(1 + K_{SA1}/A_1)$	$(1/k_2)(1 + A_2/K_{A2}) + (1/g_1)(1 + K_{SA1}/A_1)$
$nR_{21} =$	$(1/k_1)(1 + A_1/K_{A1}) + 1/g_2$	$1/k_1 + (1/g_2)(1 + K_{SA2}/A_2)$	$(1/k_1)(1 + A_1/K_{A1}) + (1/g_2)(1 + K_{SA2}/A_2)$
$nR_{ee} =$	$1/g_1 + 1/g_2$	$(1/g_1)(1 + K_{SA1}/A_1) + (1/g_2)(1 + K_{SA2}/A_2)$	$(1/g_1)(1 + K_{SA1}/A_1) + (1/g_2)(1 + K_{SA2}/A_2)$
$nR_{00} =$	$(1/k_1)(1 + A_1/K_{A1}) + (1/k_2)(1 + A_2/K_{A2})$	$1/k_1 + 1/k_2$	$(1/k_1)(1 + A_1/K_{A1}) + (1/k_2)(1 + A_2/K_{A2})$
$K =$	$(k_1/g_1)K_{A1}K_{AS1}$	$(k_1/g_1)K_{S1}K_{SA1}$	$(k_1/g_1)K_{A1}K_{AS1} = (k_2/g_2)K_{A2}K_{AS2}$
$=$	$(k_2/g_2)K_{A2}K_{AS2}$	$(k_2/g_2)K_{S2}K_{SA2}$	$(k_1/g_1)K_{S1}K_{SA1} = (k_2/g_2)K_{S2}K_{SA2}$
Constraint:	Preceding equation	Preceding equation	Preceding equation, which includes
			$K_{A1}K_{AS1} = K_{S1}K_{SA1}$
			$K_{A2}K_{AS2} = K_{S2}K_{SA2}$

[a] Where $R_{00} + R_{ee} = R_{12} + R_{21}$; n, total number of cotransporters per unit area of membrane. Terms in k, g, and K are defined in Fig. 5.6. For the flux of A (e.g., $v_{1\to2}^A$), interchange A and S everywhere in the table. (The solutions given are for the situation where the transport steps are rate-limiting.)

[b] From Stein (1986).

5.3.4. Stoichiometry of Cotransport

It need not be the case that only one molecule of the driving or driven substrate binds to the carrier to form the complex that reorients across the membrane. Just as we found for the antiporters, so with these symporters the stoichiometry of binding of the cosubstrates determines the "intensity of concentration" that is reached. One can easily show that if s molecules of substrate S and a molecules of substrate A bind simultaneously to the carrier to form EA_aS_s, the law of cotransport at the steady state will be

$$(S_1/S_2)^s = (A_2/A_1)^a \tag{5.6}$$

Thus a 10-fold ratio of the driving substrate A will bring about a $10^{a/s}$-fold ratio of the driven substrate S. The body makes important use of this phenomenon, as one might expect.

As we mentioned above, the kidney tubules contain active transport systems that reabsorb sugars and amino acids that have been filtered from the blood into the fluids from which the urine will be formed. It has been shown that glucose reabsorption takes place on two different active transporters. In the first encountered portions of the kidney, where the urinary fluid has just recently been filtered from the blood, the glucose concentration is high, similar to that in the blood. Here, therefore, large amounts of sugar have to be reabsorbed up a small concentration gradient. The intensity of active transport need not be high but the capacity for transport needs to be high. Cotransport occurs by the model of Fig. 5.6, with a 1:1 stoichiometry between sugar and sodium. By the time the urinary fluid has reached the further portions of the kidney, however, much of the sugar has been reabsorbed and its concentration within the tubule is low. It has to be absorbed up a high concentration gradient and at low capacity. James Turner and Aryeh Moran found that in this portion of the kidney, cotransport uses a system with a stoichiometry of 2 sodium ions to 1 glucose molecule. A 10-fold concentration ratio of sodium (lumen to cell) will [see Eq. (5.6)] produce a 100-fold ratio of sugar (cell to lumen) and the last traces of sugar can then be scavenged out of the tubule and returned to the body. Active transport of each molecule of sugar needs the simultaneous uptake into the cell of two ions of sodium, which will then have to be pumped out of it by the sodium pump (Section 6.1). The amount of energy obtained by the oxidation of this glucose is well in excess of the extra investment of energy used in sodium pumping.

5.3.5. Electrogenic Aspects of Cotransport: The Equilibrium Potential of a Cotransport System

We saw in Fig. 5.9 that the cotransport of glucose and sodium ions is accompanied by the movement of electrical charge, so that the rate of this cotransport can be measured as an electric current. We saw for transport on the simple carrier (Section 4.7) and for transport by an antiporter (Section 5.2.1) that there will be a mutual effect of the rate of transport and the electrical potential across the membrane. The same occurs in a cotransport system for which there is a net transport of charge across a membrane during a complete cycle of transport. Such a cotransport will be electrogenic.

We were able to define an equilibrium potential in the case of a simple ion that is capable of moving across a membrane. This was given in Eq. (2.5). In the same way, we can define an equilibrium potential for an ion moving on a cotransporter, where the flow of the ion is *coupled* to the flow of the cosubstrate. Clearly, the concentrations of both the ion and its cosubstrate must appear in the definition of the equilibrium potential. (For simplicity, we shall assume that A is charged, whereas S is uncharged.)

To define the equilibrium potential, we make use of the fundamental equation for the free energy, Eq. (2.3). At equilibrium, the free energy for the transfer of the uncharged cosubstrate from side 1 to side 2, given by $RT \ln(S_2/S_1)$, must be equal and opposite to that for the transfer of the charged cosubstrate, given by $RT \ln(A_2/A_1) + ZF \Delta\psi$.
Thus,

$$RT \ln(S_2/S_1) = - RT \ln(A_2/A_1) + ZF \Delta\psi \qquad (5.7)$$

giving

$$\Delta\psi = RT/ZF \ln[(S_2/S_1)(A_2/A_1)] \qquad (5.8)$$

Box 5.4. An Electrogenic Symporter with Stoichiometry Unequal to Unity

Clearly, Eq. (5.5) is a special case of Eq. (5.8) for the situation where the membrane potential is held at zero, and the electrogenicity of the cotransporter cycle is, therefore, of no significance. Equation (5.8) can be readily modified to take into account the possibility that the cotransporter's stoichiometry is not 1:1. The concentration ratios in Eq. (5.8) must be multiplied by the appropriate valence factor, as must the whole expression on the right of Eq. (5.8). If a ions of the cotransported substrate A and s molecules of S take part in a transport event, we rewrite Eq. (5.8) as

$$\Delta\psi = (1/a) RT/ZF \ln[(S_2/S_1)^s (A_2/A_1)^a] \qquad (B5.4.1)$$

This is the electrogenic analog of Eq. (5.6).

Equation (5.8) is very important in helping us to understand how the potential across the membrane of the living cell will affect the cotransport systems found in it. Equation (5.8) defines the equilibrium potential for the cotransporter, that is, the potential that the cotransporter can itself build up across a membrane in the absence of any short-circuiting (see Section 4.7), or the potential at which cotransport will cease if a potential is imposed by some external system. We shall see in Section 5.3.6 that we can often characterize cotransport systems by measuring the membrane potential that they establish.

The membrane potential that a cotransporter builds up can be directly measured by the experimental procedures that were described briefly in Section 3.4.2 (Figs. 3.10 and 3.11) using microelectrodes or potential-sensitive dyes. In the following sections, where we discuss some individual cotransporters, we shall consider examples of these measurements.

5.3.6. Some Individual Cotransporters Described

A few of the many cotransporter systems that have been described are listed in Table 5.2. We discuss three of them in detail.

5.3.6.1. THE LACTOSE AND MELIBIOSE SYMPORTERS OF *ESCHERICHIA COLI*

When the bacterium, *E. coli*, is starved of glucose and given instead lactose, a set of metabolic systems is induced that enables the bacterium to feed and grow on the new substrate. The exploration of this induction process led to many of the insights that shaped modern molecular biology. One particular aspect of the whole process has been just as important in developing our understanding of active transport. One of the proteins induced when such bacteria are fed lactose is *lactose permease*, a protein that is inserted into the bacterial membrane within minutes after induction of lactose metabolism and that has the capacity of allowing its substrate, lactose, to cross the plasma membrane of the bacterium, and of concentrating it within the cell. (The term "permease," with its suffix "-ase" denoting an enzyme, was introduced many years ago when it was thought that active transport was brought about by an enzyme.) The permease can transport many β-galactoside sugars. Lactose is its natural substrate. Certain of the synthetic substrates cannot be metabolized in the cell and accumulate within it to concentration gradients of over 1000. Active transport of the β-galactosides requires energy, produced by the metabolism of the cell. A separate symporter carries the disaccharide melibiose across the *E. coli* membrane. The melibiose symporter can transport either α- or β-galactosides. As we show in Box 5.5, the lactose and melibiose systems have been studied by the methods of molecular biology.

(*Text continues on page 207*)

TABLE 5.2

Some Examples of Cotransporter Systems[a]

Tissue or cell	Cosubstrates	Comments
	Inorganic ions	
Kidney	Na^+, Cl^-	Inhibited by furosemide
Gallbladder	Na^+, Cl^-	
Intestine	Na^+, Cl^-	Inhibited by furosemide
Erythrocytes	Na^+, K^+, Cl^-	Inhibited by furosemide; in birds, activated by phosphorylation
Ascites tumor cells	Na^+, K^+, Cl^-	Inhibited by furosemide
Kidney	Na^+, K^+, Cl^-	Inhibited by furosemide
Kidney; intestine	Na^+, PO_4^{3-}	Stimulated by parathyroid hormone, vitamin D
Yeast	H^+, PO_4^{3-}	$2:3$ $H^+:PO_4^{3-}$
Yeast	H^+, SO_4^{2-}	
	Sugars	
E. coli	Na^+, melibiose	Either H^+ or Na^+
Kidney; intestine	Na^+, glucose	Low affinity for sugars; well inhibited by phloridzin, $1:1$ $Na^+:Glu$
Kidney; intestine	Na^+, glucose	High affinity for sugars; less inhibited by phloridzin, $2:1$ $Na^+:Glu$
Bacteria	H^+, many different sugars	
Yeast	H^+, many different sugars	
	Amino acids	
Erythrocytes	Na^+, many different amino acids	(see Section 4.8.2) Separate systems
Kidney; intestine	Na^+, many different amino acids	Separate systems
Ascites tumor cells	Na^+, many different amino acids	Separate systems
Bacteria	H^+, many different amino acids	
Yeast	H^+, many different amino acids	
	Miscellaneous organic compounds	
Intestine	Na^+, cholate derivatives	
Kidney	Na^+, cholate derivatives	
Neuron endings	Na^+, choline	
Platelets	Na^+, Cl^-, serotonin	Antiport with K^+
E. coli	H^+, vitamin B_{12}	

[a] A few of the many symporters that have been studied. A more complete table is found in Stein (1986, p. 426–439).

Box 5.5 Molecular Biology of Two Bacterial Symporters

The Melibiose Transporter

The melibiose transporter of *E. coli* carries β-galactosides in strict symport with Na^+, but α-galactosides with either Na^+ or H^+. The molecule has been cloned and sequenced (Fig. 5.10). It consists of 469 amino acids and has a molecular mass of 52,029 Da. From its hydropathy plot (Section 1.4), the protein appears to cross the membrane 11 times.

Thomas Wilson and colleagues have studied the melibiose symporter, using molecular biological techniques that are interesting in themselves and that have delineated the substrate-binding region of the molecule. They first constructed a plasmid (a bacterial "minichromosome"; see Box 3.3) into which was inserted the gene for the melibiose symporter. This plasmid was then introduced into a strain of *E. coli* that is defective in an enzyme that repairs DNA. As a result of this defect, *all* genes in this strain (including those introduced artificially as plasmids) are subject to a very high rate of mutation, about 1000-fold more than normal. Bacteria containing genes so mutated were cultivated on a medium that contained melibiose as the sole source of energy but also contained TMG (methyl-β-D-thiogalactopyranoside), an effective competitive inhibitor of the melibiose transporter. Thus only bacteria containing mutated transporters with a weakened affinity for this inhibitor would be able to take up melibiose and survive to produce colonies. In this way 70 mutants of the transporter were identified and these were then sequenced and the position of the mutation determined. In parallel, transport studies were undertaken of the mutants and K_m and V_{max} for melibiose transport were measured.

Figure 5.10 shows the sites on the protein at which, in this way, mutations were found that affected the affinity of the transporter for its substrate. In the inset to Fig. 5.10 these mutation sites are indicated by black dots. The mutations are clustered in four regions of the protein, although the technique used for mutagenesis would have produced mutations at all sites on the protein! One can assume that these four regions of the protein are close to one another when the polypeptide chain of the protein is folded into a three-dimensional structure, and that this cluster of residues forms the substrate-binding center of the transporter. Most interestingly, all of these mutated transporters (although selected for their poor affinity for the sugar) also have a lowered affinity for lithium. Now, lithium seems to bind to the transporter at its

(Continued)

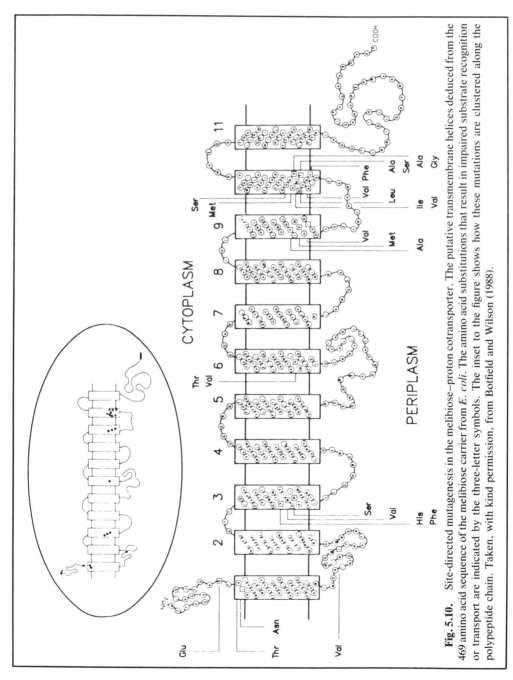

Fig. 5.10.   Site-directed mutagenesis in the melibiose–proton cotransporter. The putative transmembrane helices deduced from the 469 amino acid sequence of the melibiose carrier from *E. coli*. The amino acid substitutions that result in impaired substrate recognition or transport are indicated by the three-letter symbols. The inset to the figure shows how these mutations are clustered along the polypeptide chain. Taken, with kind permission, from Botfield and Wilson (1988).

(Continued)

ion-binding site. Indeed, lithium at low concentrations can substitute for sodium as the symported ion, although at high concentrations it inhibits.

In addition, a mutation at position 122 (the left-most site with label Ser in Fig. 5.10) converts the transporter from the natural form that can symport α-galactosides using either H^+ or Na^+ to one that has an absolute requirement for Na^+ or Li^+.

These results with cations suggest that the ion-binding site is situated close to the sugar-binding site.

The Lactose Permease

The cation-binding site, this time of the lactose permease of *E. coli*, has been explored by Ronald Kaback and colleagues. Studies with specific chemical reagents had suggested that a histidine residue of the protein was intimately involved in the coupling of lactose and proton transport on this symporter. Using the method of site-directed mutagenesis (see Box 3.3), all four histidines in the protein were separately mutated and replaced by other amino acids. Only the histidine residue in position 322 (located in helix 10 of Fig. 5.11) proved to be essential for the full functioning of the transporter. When the neighboring glutamate residue (Glu-325) was also replaced (by alanine), an interesting result was found. The constructed protein was shown to be unable to *concentrate* lactose, but it was able to *exchange* radioactively labeled lactose with the unlabeled sugar! The defect in transport proved to be in the inability of the lactose-free carrier to return across the membrane. This defect is associated with its inability to off-load the bound proton. (See the discussion in the text regarding Figs. 5.12 and 5.13.) Thus His-322, Glu-325, and also, apparently, Arg-302 are involved in that part of the overall symport that is most directly concerned in the movement of the proton.

The results throw much light on the mechanism of symport and bring us closer to an understanding of how these active transporters work.

The gene coding for the lactose permease has been cloned and sequenced. The protein is composed of 417 amino acids and has a molecular mass of 46,504 Da. Its sequence suggests that it spans the cell membrane about 12 times (Fig. 5.11). Kaback and colleagues have succeeded in isolating the protein itself from the cell membrane and reconstituting it as an active cotransporter in artificial phospholipid vesicles. Using such

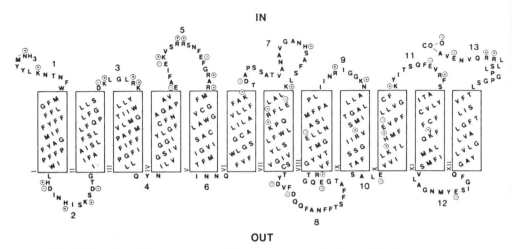

Fig. 5.11.   Possible arrangement of the transmembrane helices of the lactose permease of *E. coli*. A model for the lactose permease based on the amino acid sequence as deduced from the cloned gene. The amino acids are symbolized by the single-letter code. Positively and negatively charged amino acids are indicated. The transmembrane helices are numbered with Roman numerals, the sequences that connect these, by Arabic numerals. Taken, with kind permission from H. R. Kaback (1988). *Annu./Rev. Physiol.* **50,** 243–256.

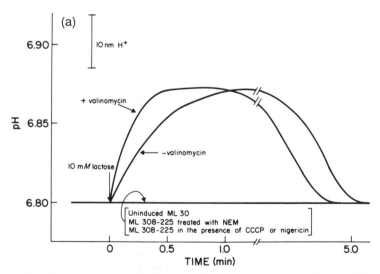

Fig. 5.12.   Lactose–proton cotransport by the lactose permease of *E. coli*. (a) Lactose-induced proton movements into *E. coli* membrane vesicles. The pH is measured in the medium, so lactose entry takes protons with it. At the first arrow, 10 mM lactose was added to the vesicles in the absence or presence of valinomycin (which short-circuits the buildup of a transmembrane potential). The horizontal line represents the data for various controls, namely, uninduced *E. coli*, or *E. coli* membranes treated with NEM, an inhibitor of the

vesicle preparations or vesicles prepared directly from the bacteria, Kaback's group reported some fascinating results that throw much light on the mechanism of action of the system. Figure 5.12(a) shows measurements of the pH *outside E. coli* vesicles in a preparation when lactose is added to the system at the time indicated by the vertical arrow.

There are three curves to consider in Fig. 5.12(a). The lowest, horizontal curve, showing no pH change in the medium, is for the control experiments in which lactose is added to noninduced bacteria, or the lactose permease is inhibited by the addition of a mercurial [*N*-ethylmaleimide (NEM)], or a molecule is added that allows protons to equilibrate rapidly

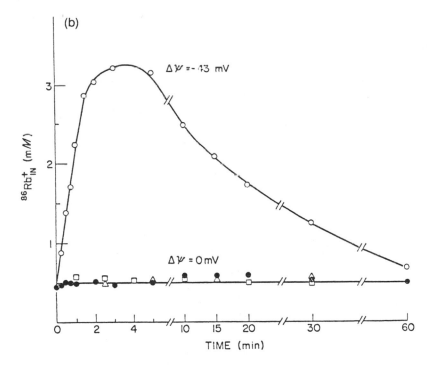

system, or with CCCP or nigericin, both proton ionophores. Reprinted with permission from L. Patel, M. L. Garcia, and H. R. Kaback (1982). *Biochemistry* **21,** 5805–5810. Copyright 1982 American Chemical Society. (b) Development of a transmembrane potential as a result of the uptake of protons in the system described in (a) above. At time zero, preloaded lactose (initially 10m*M*) was allowed to flow *out* of the membrane vesicles and the uptake of rubidium (driven by the potential) was measured. The horizontal line represents control experiments in which lactose was added in addition outside the cells (so that there was no net flow), or CCCP was added or the inhibitor PCMBS. Reprinted with permission from M. L. Garcia *et al.* (1983). *Biochemistry* **22,** 2524–2531. Copyright 1983 American Chemical Society.

across the membrane [an "uncoupler" (see Section 6.6.1) such as car-
bonyl cyanide m-chlorophenylhydrazone (CCCP)]. The steepest curve
shows the rapid pH change when lactose is added to membranes in the
present of valinomycin. [As we saw in Chapter 4, valinomycin is a carrier
of potassium ions. It thus acts to short-circuit any potential induced by
the lactose cotransporter itself. The rate of pH change is correspondingly
greater than when valinomycin is omitted (the middle curve in Fig.
5.12(a)), where a pH flow still occurs but is slower than with valinomycin
present].

What does Fig. 5.12(a) suggest? The external pH rises when lactose is
taken up by the cells. This means that protons are disappearing from the
external medium. Thus lactose is taken up into the vesicles in cotransport
with protons! Indeed, direct measurements of the simultaneous uptake of
lactose and of protons showed a 1:1 stoichiometry. One can show, too,
that the cotransport of lactose and proton is electrogenic. Consider the
experiment depicted in Fig. 5.12(b) (also from Kaback's laboratory). In
this experiment, use is made of the fact that the bacterial cell membranes
contain potassium channels. Measurement of the flux of potassium (or of
rubidium, an easier to handle analog of potassium) allows a determination
of the transmembrane potential. Lactose was allowed to flow *out* of membrane vesicles into a medium containing radioactive rubidium. The uptake
of rubidium into the vesicles is plotted on the figure as a function of time
after addition of the lactose. The control experiment was once again the
addition of a mercurial to inhibit the permease or of CCCP to prevent the
buildup of a pH gradient (the horizontal line in Fig. 5.12(b)). The steep
uptake of rubidium occurred only when lactose efflux took place, accompanied by a net efflux of protons. The latter flux caused a membrane
potential to be built up, inside negative (as we discussed in the previous
section), and this pulled labeled rubidium into the cell. As the protons
gradually leaked back into the vesicle, the membrane potential was dissipated and the flow of rubidium reversed.

Figure 5.13 shows additional data from Kaback's laboratory illustrating exactly how this
cotransport occurs. Figure 5.13(a) shows a zero trans experiment in which lactose efflux was
studied as a function of the external pH. In the control experiment [the uppermost curve of
Fig. 5.13(a)] vesicles are treated with a mercurial inhibitor of transport. With no inhibitor, as
the pH in the *external* medium is successively raised, the rate of efflux of lactose increases.
To understand this result, return to Fig. 5.6, where kinetic models for cotransport were
presented. Let S in these figures be lactose and A be proton, while side 1 will be the vesicle
interior. Why should decreasing the concentration of proton (A) at side 2 increase the rate of
lactose (S) flow from side 1 to side 2? A little reflection will convince the reader that scheme
(b) in Fig. 5.6 is consistent with the data for lactose cotransport. We have here, indeed, a
case of trans inhibition, similar to that described in Section 5.3.3. In Fig. 5.6(b), as the
concentration of A (here proton) at side 2 is increased, the cotransporter is trapped in the
form EA_2 and cannot return to side 1 to bring about further efflux of lactose. Thus as the pH

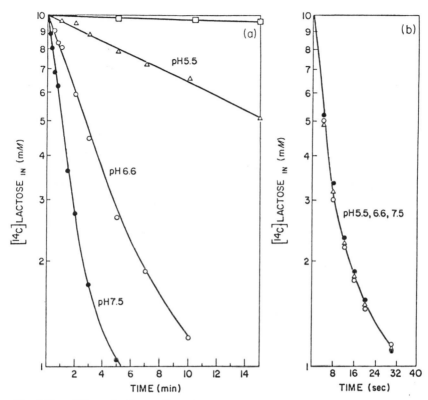

Fig. 5.13. Effect of pH on efflux of labeled lactose from *E. coli* membrane vesicles. Efflux of preloaded lactose (initially 10 m*M*) from membrane vesicles as in Fig. 5.12(b). Part (a) shows a zero trans efflux at three different external pH values. (The uppermost curve is a control treated with an inhibitor of transport.) Part (b) shows data for an infinite trans experiment (see section 4.2.4) (△, pH 5.5; ○, pH 6.6; ●, pH 7.5.) Reprinted with permission from M. L. Garcia *et al.* (1983). *Biochemistry* **22**, 2524–2531. Copyright 1983 American Chemical Society.

is increased (and the proton concentration falls), more and more cotransporter will be liberated as E_2, to return to side 1 for a further cycle of cotransport. This conclusion is confirmed by the data depicted in Fig. 5.13(b). This represents the efflux of lactose in an *infinite-trans* experiment, i.e., in which the concentration of lactose at side 2 is very high and one measures the exchange of labeled lactose for unlabeled. Changes in the external pH now have no effect. Returning to Fig. 5.6(b) will convince the reader that in these circumstances, EA_2 is never formed and hence cannot break down to E_2 at any pH. Exchange can occur when *S* dissociates from EAS_2, but its place is immediately taken by *unlabeled S* with a return of cotransporter to side 1 for a further cycle of transport.

Thus the experiments of Fig. 5.13 prove that the order of association of the cotransported substrates at the external face of the cell is proton first, lactose second. The system is a typical cotransporter, in which lactose uptake is driven by the inwardly directed gradient of protons, which is itself built up by primary transport linked to the metabolism of the cell.

5.3.6.2.   ACCUMULATION OF A NEUROTRANSMITTER
           IN STORAGE GRANULES

The neurotransmitter serotonin (or 5-hydroxytryptamine) is an important natural pharmacologically active substance, found in the blood in storage granules within the platelets (and also in nerve cell endings and in mast cells). Its concentrations within these storage granules can reach very high levels: 0.6 M has been measured when the extracellular concentration of the substance was only 10^{-5} M, a concentration gradient of 6 × 10^4! Research has shown that there are two quite distinct mechanisms for serotonin accumulation, the two working in tandem. One is present in the cytoplasmic membrane of the platelet and is a typical cotransporter bringing about a 100-fold concentration of serotonin. The other system is at the membrane of the granule itself. This is an antiporter and concentrates the serotonin an additional 600-fold. We discuss the two systems in turn.

Vesicles isolated from platelet plasma membranes accumulate serotonin, provided that a gradient of sodium is present. The intact platelet cell develops, however, only a fourfold concentration gradient for sodium and, therefore, this gradient in itself is far from sufficient to bring about the required concentration of serotonin. One must also take into account the effect of the membrane potential, which is an additional source of energy driving the sodium into the cell and also driving in the positively charged serotonin. How does the system harness the membrane potential?

The intriguing observation was made that the system is nonelectrogenic. Cotransport of Na^+ and serotonin (both positively charged) should transport *two* charges across the membrane and hence be decidedly electrogenic. What is neutralizing these two charges? Clearly, either cations are being coupled, in antiport, and move inward as serotonin and Na^+ move outward, or anions accompany these two cations as they enter the cell, in cotransport. It was found that both of these situations are operative. Serotonin uptake requires cotransport of sodium but is also coupled, on the same carrier, to antiport of potassium ions. In addition, chloride is an obligatory substrate for serotonin uptake, as it is cotransported into the cell together with it and sodium. Potassium neutralises the electrogenic effect of one of the two inwardly transported cations, chloride neutralizes the other, and thus the system as a whole is nonelectrogenic. Transport of any ion on the system cannot be affected by the prevailing membrane potential (itself, or course, a Donnan potential; see Section 3.7). Thus, the system can harness the fourfold gradient of sodium ions in the cell, its 10-fold gradient of potassium ions and the 10-fold gradient of chloride to build up the serotonin concentration in the cytoplasm to over 100-fold.

Other neurotransmitters also move by cotransport with sodium ions and chloride ions and by antiport with potassium ions; different combinations of these are found for different neurotransmitters.

We are only halfway, however, to an explanation of how serotonin is accumulated in the storage granules to a level 6×10^4 times that in the extracellular fluid. Work with the isolated granules has shown that these contain a proton–serotonin antiporter. This exchanges two protons for one molecule of serotonin. This system, which is obviously electrogenic, is sensitive not only to the concentration gradient of protons but also to the electrical potential that prevails across the membrane between granule interior and the cytoplasm. This potential is about 40 mV (inside negative) which according to Eq. (2.5), should be sufficient to bring about a fivefold concentration ratio of serotonin. Finally, there is a pH difference of at least 2, inside low, across the membrane of the storage granule. This should give a concentration ratio of 100-fold for a $1:1$ stoichiometry of proton to serotonin, and 100×100 for the $2:1$ stoichiometry that the system manifests. There is clearly more than enough free energy available to provide for the granule's ability to concentrate serotonin. (The pH gradient and the prevailing membrane potential are themselves maintained by a *primary* active transport of protons, coupled to the hydrolysis of ATP. We will discuss the mechanism of action of such systems in detail in Chapter 6.)

5.3.6.3. THE UBIQUITOUS $Na^+ + K^+ + 2Cl^-$ COTRANSPORTER

Many cells of the body have been shown to possess a cotransporter of the three major intracellular ions: sodium, potassium, and chloride. We shall see in Chapter 7 that this system is present in those *epithelia* that have the capacity to pump salt across them. The cotransport system admits salt at one side of the epithelium, and the sodium pump extrudes it at the other. A major salt-transporting organ such as the kidney is replete with these cotransporters. The system is also widely used for the short-term regulation of the volume of cells, which, too, we shall discuss in Chapter 7. Here we consider the mechanism of action of the cotransporter, that is, how it behaves as a transporter of its three cosubstrates.

Most important for the progress of research on this system was the finding that it is inhibited rather specifically by the drug furosemide (and its analog bumetanide). This is one of the "loop diuretics" that encourage loss of fluid through the kidney ("diuresis") by blocking salt movement across the tubules in the loop of Henle. Indeed, its action on the $Na^+ + K^+ + Cl^-$ cotransporter accounts for the diuretic action of furosemide. By

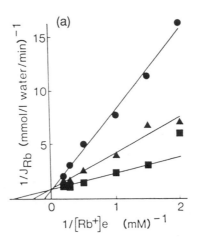

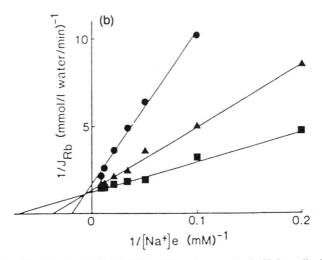

Fig. 5.14.   Kinetics of the $Na^+ + K^+ + 2Cl^-$ cotransporter in HeLa cells. Parts (a)–(d) show reciprocal plots of the ouabain-insensitive, furosemide-sensitive rubidium fluxes into cultured HeLa cells (Rb^+ is a potassium analog in this system). Part (a) shows the data at three fixed concentrations of sodium (10, 30, 140 mM, bottom to top) with rubidium varying; (b) is as above but with Na^+ varying and rubidium fixed at 1, 2, and 5 mM (top to bottom); (c) as (a) but with Cl^- fixed at 45, 60, and 140 mM (top to bottom); (d) Rb^+ entry measured as a function of the chloride concentration at fixed Rb^+ and Na^+. Taken, with kind permission, from H. Miyamoto *et al.* (1986). *J. Membr. Biol.* **92**, 135–150. (*Figure continues*)

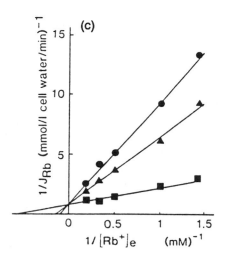

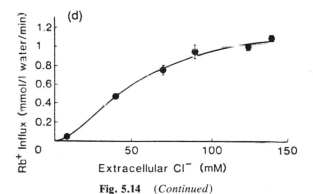

Fig. 5.14   (*Continued*)

measuring the furosemide-sensitive component of an ion flux, one can estimate how much of the flux is carried on the cotransporter.

Consider Fig. 5.14(a), which depicts data obtained on the uptake of rubidium into HeLa cells as a function of the rubidium concentration at three different concentrations of sodium. (HeLa cells were isolated many years ago from a tumorous growth in a human patient, Helen Lane, and subcultured since then.) In this system, rubidium can once again take the place of its analog potassium. Clearly, at all three concentrations of sodium, uptake of rubidium reaches the same maximum velocity, but K_m is decreased, i.e., the apparent affinity increased, with increasing sodium concentration. This is the K kinetics that we discussed in connection with Fig. 5.9. Consider, again, the models for cotransport depicted in Fig. 5.6.

Which model can account for the data of Fig. 5.14(a)? Recalling our previous discussion on the V and K kinetics (Section 5.3.2), we can conclude that it must be sodium that binds first to the cotransporter, rubidium (or potassium) second. This is confirmed by the data of Fig. 5.14(b). Here Rb uptake is studied as a function of sodium concentration at three fixed concentrations of rubidium. The kinetics are as follows: changes in both V_{max} and K_m are to be seen, V_{max} increasing and K_m decreasing as the concentration of rubidium is raised. As discussed in Section 5.3.2, however, this is just what Fig. 5.6 would predict if sodium were to bind first and rubidium (or potassium) second to the cotransporter.

What of the chloride ions? Figure 5.14(c) shows that altering the rubidium concentrations at three fixed concentrations of chloride gives V kinetics. Thus, from our discussion in Section 5.3.2, the order of addition must be random for rubidium and chloride. Finally, Fig. 5.14(d) shows rubidium uptake as a function of the chloride ion concentration. The curve is not a simple Michaelis–Menten function but is sigmoidal, consistent with the involvement of *two* chloride ions in the cotransport process. Other experiments show that one chloride ion is associated with the sodium-binding site and binds after the sodium, the other with the rubidium site and, as we saw from Fig. 5.14(c), binds in random order with the rubidium.

Overall, the kinetics show that the order of addition is sodium, then chloride, and then rubidium and/or chloride. The final complex is electrically neutral, hence the system as a whole is nonelectrogenic, neither producing an electric current during transport, nor being affected by the prevailing membrane potential. In addition, after each stage of cation binding, electroneutrality is maintained.

Although the data of Fig. 5.14 for the HeLa cell system are perfectly clear and unambiguous, it should be pointed out that in other $Na^+ + K^+ + Cl^-$ cotransport systems, other kinetic models have been suggested. For instance, the system in mouse 3T3 cells seems to bind all its substrates in random fashion.

The binding of the inhibitor bumetanide to cell membranes containing the $Na^+ + K^+ + 2Cl^-$ cotransporter is strongly affected by the presence of the cotransported ions. Sodium and potassium ions are both required for the binding of the drug and the concentrations at which they half-stimulate binding are close to the K_m values found for their cotransport. Chloride ions at lower concentrations stimulate the binding of the drug, but at higher concentrations they inhibit. This fact suggests that the drug binds to the site at which the second chloride would otherwise bind—and thus blocks the cotransport. A radioactively labeled analog of bumetanide binds to a 150-kDa polypeptide in cell membranes that contain the symporter and this polypeptide may be the symporter. Cloning of the gene of the symporter should in the future provide a firm identification of the protein.

A most important aspect of the behavior of many of these $Na^+ + K^+ + Cl^-$ cotransporters is that their activity can be *regulated* by physiologic factors. This important subject is treated in Chapter 7.

Suggested Readings

General

Stein, W. D. (1986). "Transport and Diffusion across Cell Membranes," Chapters 4 and 5. Academic Press, Orlando, Florida.

Countertransport

Frohlich, O. (1989). Antiporters. *Curr. Opin. Cell Biol.* **1,** 729–734.
Rosenberg, T., and Wilbrandt, W. (1957). Uphill transport induced by counter-flow. *J. Gen. Physiol.* **41,** 289–296.
Widdas, W. F. (1952). Inability of diffusion to account for placental glucose transfer in the sheep and consideration of the kinetics of a possible carrier transfer. *J. Physiol. (London)* **118,** 23–39.

Kinetics of Antiport

Aronson, P.S. (1985). Kinetic properties of the plasma membrane Na^+-H^+ exchanger. *Annu. Rev. Physiol.* **47,** 546–560.

ADP/ATP Exchange

Klingenberg, M. (1980). The ADP-ATP translocation in mitochondria, a membrane potential controlled transport. *J. Membr. Biol.* **56,** 97–105.

Na^+/H^+ Antiporter

Aronson, P. S., and Igarashi, P. (1986). Molecular properties and physiological roles of the renal Na^+-H^+ exchanger. *Curr. Top. Membr. Transp.* **26,** 57–75.
Grinstein, S. *et al.* (1989). Activation of sodium–hydrogen exchange by mitogens. *Curr. Top. Membr. Transp.* **34,** 331–344.
Villereal, M. L. (1986). Na^+-H^+ and Na^+-Ca^{2+} exchange in activated cells. *Curr. Top. Membr. Transp.* **27,** 55–88.

Growth Factors

Rozengurt, E., and Mendoza, S. A. (1986). Early stimulation of Na^+-H^+ antiport, Na^+-K^+

pump activity, and Ca^{2+} fluxes in fibroblast mitogenesis. *Curr. Top. Membr. Transp.* **27**, 163–191.

Villereal, M. L. *et al.* (1986). Mechanisms of growth factor stimulation of Na^+–H^+ exchange in cultured fibroblasts. *Curr. Top. Membr. Transp.* **26**, 175–192.

Na^+–Ca^{2+} Antiporter

Blaustein, M. P. (1989). Sodium–calcium exchange in cardiac, smooth, and skeletal muscles: Key to contractility. *Curr. Top. Membr. Transp.* **34**, 289–330.

Mullins, L. J. (1984). An electrogenic saga: Consequences of sodium-calcium exchange in cardiac muscle. *In* "Electrogenic Transport: Fundamental Principles and Physiological Implications" (M. P. Blaustein and M. Lieberman, eds.), pp. 161–179. Raven Press, New York.

Electrogenicity

Lieberman, M. *et al.* (1984). Physiologic criteria for electrogenic transport in tissue-cultured heart cells. *In* "Electrogenic Transport: Fundamental Principles and Physiological Implications" (M. P. Blaustein and M. Lieberman, eds.), pp. 181–191. Raven Press, New York.

Cotransport Systems

Crane, R. K. (1977). The gradient hypothesis and other models of carrier-mediated active transport. *Rev. Physiol. Biochem. Pharmacol.* **78**, 99–159.

Kimmich, G. A. (1990). Membrane potentials and the mechanism of intestinal Na^+-dependent sugar transport. *J. Membr. Biol.* **114**, 1–27.

Stein, W. D. (1989). The cotransport systems. *Curr. Opin. Cell Biol.* **1**, 739–745.

Ussing, H. H. (1954). Active transport of inorganic ions. *Symp. Soc. Exp. Biol.* **8**, 407–422.

Wiseman, G. (1953). Absorption of amino acids using an *in vitro* technique. *J. Physiol. (London)* **120**, 63–72.

Molecular Biology of the Sodium–Glucose Symporter

Hediger, M. A. *et al.* (1987). Expression cloning and cDNA sequencing of the Na^+/glucose co-transporter. *Nature (London)* **330**, 379–381.

Sanders, D., Hansen, U.-P., Gradmann, D., and Slayman, C. L. (1984). Generalised kinetic analysis of ion-driven cotransport systems: A unified interpretation of selective ionic effects on Michaelis parameters. *J. Membr. Biol.* **77**, 123–152.

Semenza, G. *et al.* (1984). Biochemistry of the Na^+, D-glucose cotransporter of the small-intestinal brush-border membrane. *Biochim. Biophys. Acta* **779**, 343–379.

Stoichiometry of Cotransport

Turner, R. J., and Moran, A. (1982a). Stoichiometric studies of the renal outer cortical brush border membrane D-glucose transporter. *J. Membr. Biol.* **67**, 73–80.

Turner, R. J., and Moran, A. (1982b). Further studies of proximal tubular brush border membrane D-glucose transport heterogeneity. *J. Membr. Biol.* **70**, 37–45.

Melibiose Transport

Botfield, M. C., and Wilson, T. H. (1988). Mutations that simultaneously alter both sugar and cation specificity in the melibiose carrier of *Escherichia coli*. *J. Biol. Chem.* **263**, 12909–12915.

Lactose Permease

Overath, P. *et al.* (1987). Lactose permease of *Escherichia coli*: Properties of mutants defective in substrate translocation. *Proc. Natl. Acad. Sci. U.S.A.* **84**, 5535–5539.

Puttner, I. B., and Kaback, H. R. (1988). *lac* permease of *Escherichia coli* containing a single histidine residue is fully functional. *Proc. Natl. Acad. Sci. U.S.A.* **85**, 1467–1471.

Amino Acid Cotransport

Eddy, A. A. (1987). The sodium gradient hypothesis of organic solvent transport with special reference to amino acids. *In* "Amino Acid Transport in Animal Cells" (D. L. Yudilevich and C. A. R. Boyd, eds.), Manchester Univ. Press, Manchester, U.K. pp. 47–86.

Ellory, J. C. (1987). Amino acid transport in mammalian red cells. *In* "Amino Acid Transport in Animal Cells" (D. L. Yudilevich and C. A. R. Boyd, eds.), pp. 106–119. Manchester Univ. Press, Manchester, U.K.

Yudilevich, D. L., and Boyd, C. A. R., eds. (1987). "Amino Acid Transport in Animal Cells." Manchester Univ. Press, Manchester, U.K.

Na$^+$–K$^+$–2Cl$^-$ Cotransporter

Haas, M., and Forbush, B. (1988). Photoaffinity labelling of a 150 kDa (Na + K + Cl)-cotransport protein from duck red cells with an analog of bumetanide. *Biochim. Biophys. Acta* **939**, 131–144.

O'Grady, S. M. *et al.* (1987). Characteristics and functions of Na–K–Cl cotransport in epithelial tissues. *Am. J. Physiol.* **253**, C177–C192.

CHAPTER 6

Primary Active Transport Systems

We saw in Chapter 5 that counter- and cotransport of many metabolites into and out of the living cell are brought about by coupling of such flows with the flow of sodium or of hydrogen ions. By these means, the electrochemical gradient of the pumped metabolite is coupled to the existing gradient of the sodium ion or proton. But how is the electrochemical gradient of the sodium or proton itself established? In this chapter we shall see that these ions are pumped out of the cell or organelle by *primary* active transport systems that use metabolic energy directly. We shall also see that the source of metabolic energy in the different primary active transporters may be the splitting of ATP, the harnessing of the flow of electrons in an oxidation–reduction reaction, or the absorption of the energy of a photon in the light-driven pumps.

We shall start our discussion by considering the sodium pump of animal cells, and continue with a consideration of several other, closely related cation pumps. Then we discuss the ion pumps of bacterial cells, mitochondria, and chloroplasts, and finally, the light-driven pumps.

6.1. THE SODIUM PUMP OF THE PLASMA MEMBRANE

6.1.1. The Function of the Sodium Pump

We all know that human red blood cells can be withdrawn from the body, stored under refrigeration for several weeks, and then transfused into a patient where they will function adequately in maintaining life processes in the recipient. After many years of patient exploration medical researchers worked out the best methods for storing blood. Blood cells freshly drawn from the body have a sodium ion concentration that is about one-tenth that of the extracellular medium, the blood plasma. In contrast, their potassium ion concentration is more than 10-fold that of the plasma. On cold storage, the cells lose potassium and gain sodium ions, until cellular and extracellular levels of these ions become almost equal, at about 15 mM for potassium and about 120 mM for sodium (see Fig. 6.1, points at time zero). If the cells are then incubated in the presence of a metabolizable substrate, sodium is ejected from the cell and potassium taken up until the ion concentrations normally present in the cell are restored (Fig. 6.1, lines marked "potassium added"). Thus cells for transfusion must be stored with an adequate supply of metabolizable substrate and then incubated before transfusion takes place. Now, the subsequent movement of both sodium and potassium ions is against their concentration gradients. Hence, *both* cannot be driven by the mutual coupling of their flows. Something other than an antiport or symport is

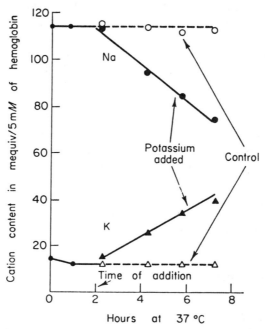

Fig. 6.1.   Sodium and potassium pumping across red cell membranes. Ion movements (sodium efflux and potassium influx) from human red blood cells stored at 2°C for 9 days and then incubated at 37°C with a metabolizable substrate (inosine). The circles show data for sodium; the triangles, for potassium. Filled symbols indicate incubations in presence of external potassium (at 21 mM); empty symbols, in absence (controls). When fresh, these cells contained about 100 mM potassium and 40 mM sodium ions. Taken with permission, from R. L. Post *et al.* (1957). *Biochim. Biophys. Acta* **25**, 118–128.

required here. We find that the system needs the direct consumption of metabolically produced ATP for the pumping of both ions. The system is thus a primary active transporter, the so-called sodium pump.

A major step forward in the study of this ion pump came from an experiment performed by the Hungarian biochemist, George Gardos. If red blood cells are swollen osmotically (see Section 2.7), pores open up in the plasma membrane. The cells lose their internal hemoglobin and, of course, their red color. They are then termed "ghosts." During the process of swelling of the cells, as the hemoglobin leaves, normally nonpermeable substances can be induced to enter the cell through the pores. Gardos loaded red cells in this way with ATP and arranged the ion concentrations of the cells so that sodium was at the same level at the two faces of the membrane, as was potassium. Gardos showed, by direct measurement, that the cells hydrolyzed the ATP as they pumped sodium out of, and pumped potassium into, the cell. Adenosine triphosphate added outside the cell could not be used; only that present at the intracellular face of the cell membrane could. Ian Glynn showed that pumping out of sodium at an appreciable rate requires the presence of potassium ions outside the cell. (Evidence for this is shown in Fig. 6.1.) Careful quantitative stoichiometric studies showed that three sodium ions are pumped out of the cell and two potassium ions are pumped in for each molecule of ATP that is split. Finally it was shown that the split products of the ATP, namely, ADP and inorganic phosphate ion (P_i), are liberated at the intracellular face of the membrane. Pumping requires that sodium and ATP be present inside the cell, potassium outside, and under these conditions hydrolysis of ATP brings about the coupled ion pumping.

Now this hydrolysis of ATP is the typical action of an ATPase enzyme. The finding that the pump splits ATP in the presence of sodium and potassium ions meant that cell membranes must contain an ATPase that is activated by sodium and potassium. Sure enough, the Danish physiologist Jens Skou soon found such an enzyme, originally isolating it from crab nerves (although it was later shown to be present in nearly all animal cell plasma membranes). This enzyme is the sodium-, potassium-activated ATPase (the Na^+,K^+-ATPase). An important advance was made by Hans Schatzmann, who showed that the enzyme could be totally, and specifically, inhibited by the drug ouabain, found in the seeds of the plant *Strophanthus*. Ouabain is one of the glycosides, used clinically today in the treatment of heart conditions. (Another of this family is digitalin, found in the foxglove plant *Digitalis*.) The identification of this powerful inhibitor meant that experiments could be easily conducted to study the role of the pump enzyme, since its action could always be blocked by ouabain.

One such experiment was performed by Robert Post. He showed that

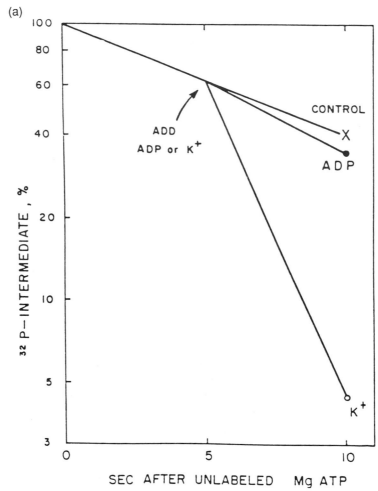

Fig. 6.2. Dephosphorylation of phosphorylated Na⁺, K⁺-ATPase. (a) Dephosphorylation of the native Na⁺, K⁺-ATPase enzyme from guinea pig kidneys at 0°C. The enzyme was
first phosphorylated from Na⁺ and radioactively labeled ATP, and then the breakdown of
the phospho intermediate visualized by adding excess unlabeled ATP. At the arrow, ADP,
K⁺ or nothing was added. (b) Dephosphorylation of the NEM-blocked enzyme. As in (a),
but the enzyme was first treated with *N*-ethylmaleimide (which, this experiment suggested,
blocks the interconversion of two conformations of the enzyme). Taken, with permission,
from Post *et al.* (1969). *J Gen Physiol* **54,** S306–S326. Reproduced by copyright permission
of The Rockefeller University Press. (*Figure continues*)

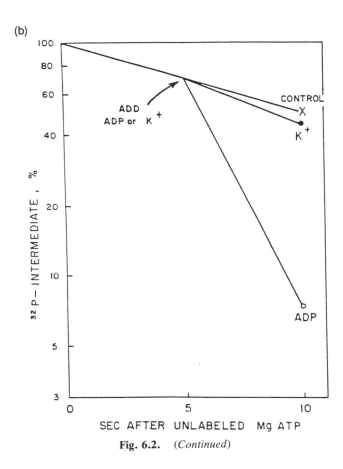

(b)

ADD
ADP or K⁺

CONTROL
X

K⁺

ADP

^{32}P – INTERMEDIATE , %

SEC AFTER UNLABELED Mg ATP

Fig. 6.2. (*Continued*)

cell membranes, isolated from dog kidney, became phosphorylated when they were incubated with ATP (radioactively labeled in the terminal phosphate residue) in the presence of sodium ions. This transfer of labeled phosphate from ATP to the membrane proteins was inhibited by ouabain and hence a function of the Na^+,K^+-ATPase. The addition of potassium to the phosphorylated membranes released from them their covalently bound phosphate, the process again being inhibited by ouabain. This meant that the ATP splitting must be taking place in two steps, each step being activated by one of the two cations—phosphorylation by sodium, dephosphorylation by potassium—to give an overall hydrolysis of ATP, activated by the combination of Na and K. Post found that treatment of kidney membranes with the sulfhydryl reagent, N-ethylmaleimide (NEM), an inhibitor of this ATPase, enabled the two steps to be functionally dissected. As Fig. 6.2 shows, phosphorylated membranes in the absence of NEM were, of course, dephosphorylated by potassium [Fig. 6.2(a)]. But if NEM was added [Fig. 6.2(b)], potassium failed to activate dephosphorylation, and only by adding ADP could the phosphate be released from the enzyme (and transferred back to reform ATP!). NEM blocked the transformation of the enzyme to the potassium-sensitive form. Post hypothesized that two conformations of the enzyme must exist. The first to form is ADP-sensitive, the second to form potassium-sensitive, while NEM paralyzes the transformation of the first to the second. These two forms of the phosphorylated pump enzyme are now known as E_1P and E_2P (P symbolizing the covalently bound phosphate residue). Whereas E_1P is ADP-sensitive, E_2P is potassium-sensitive. Identification of these two forms enabled Post, together with Albers, to propose a scheme for linking the biochemical events of ATP splitting and the transport events of sodium–potassium exchange. An updated version of their scheme, known as the *Albers–Post model*, is depicted in Fig. 6.3. In this scheme, E_1 and E_2 are two forms of the pump enzyme in which the cation-binding sites face the intracellular surface and the extracellular surface of the membrane, respectively. In the presence of *intracellular* sodium ions (three in number) and ATP, E_1 is phosphorylated, releasing ADP and yielding E_1PNa. This then undergoes a conformation change to yield E_2PNa, the

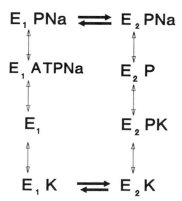

Fig. 6.3.   Kinetic scheme for Na$^+$, K$^+$-ATPase. The Albers–Post scheme for the action of the sodium–potassium pump. The symbol E is the pump enzyme; Na and K, the cations; ATP and P, the phospholigands. Symbols 1 and 2 represent the conformations with sodium-binding sites facing the cytoplasm and the extracellular phases, respectively. The single-headed arrows represent conformational changes of the enzyme. The double-headed arrows represent pump-ligand association/dissociation equilibria.

form in which the cation-binding sites face the extracellular surface. The three sodium ions dissociate from this form to yield E_2P and two potassium ions from the *extracellular* surface take their place, yielding E_2PK. This form now undergoes dephosphorylation to E_2K, finally interconverting to E_1K. The potassium ions, now bound to a site that is accessible from the intracellular surface, dissociate and release free E_1 once more, to restart the whole cycle. In a single cycle, three sodium ions are transferred from the intracellular to the extracellular medium, two potassium ions are transferred in the opposite direction, while a molecule of ATP is split to ADP and P_i. The movements of sodium and potassium and the splitting of ATP are strictly coupled, although we do not know how the coupling is controlled. This remains as a crucial problem for future investigators to solve.

The two forms E_1 and E_2, whether phosphorylated or not, differ in conformation. Thus the two behave differently toward proteolytic digestion. Peter Jorgensen from Denmark showed that the former is split in two places by chymotrypsin, the latter only in a third site, coinciding with neither of the first two. Steven Karlish showed that the two forms differed with respect to fluorescent properties, a finding that has enabled investigators to undertake many studies on the rates of interconversion between the two conformations. We now know that the scheme of Fig. 6.3 is a simplification of the true picture since a major, intermediate form that is bound to potassium has its cation-binding and cation-releasing sites accessible at neither side of the membrane. The potassium in this form is said to be "occluded" within the enzyme. Sodium can also be "occluded" in E_1PNa, the left-hand form of the two sodium-bound forms shown in Fig. 6.3.

Interestingly, the Na^+,K^+-ATPase recognizes an analog of phosphate, the anion vanadate, at the site on the enzyme that normally binds phosphate. This vanadate binds to the enzyme relatively well, but its dissociation from the enzyme, under the influence of potassium, is exceedingly slow. Hence, once bound, it remains bound to the enzyme, blocking the phosphate-binding center, and inhibiting the enzyme from taking part in any further cycle of phosphorylation and dephosphorylation. It seems that all the cation-activated ATPases that demonstrate an enzyme intermediate with phosphate bound to the enzyme can be inhibited by vanadate in such a nearly irreversible reaction. The members of this family of enzymes, of which we will discuss three other members in the following sections, are referred to as the *vanadate-inhibitable* ATPases, or following Peter Pedersen and Ernesto Carafoli's suggestion: "P-ATPases"—since they form a phosphorylated intermediate during their action.

Box 6.1. Additional Modes of Action of the Sodium Pump

Other major features of the pump enzyme that can be inferred directly from Fig. 6.3 concern the ability of the enzyme to catalyze the exchange of intracellular and extracellular sodium, and of intracellular and extracellular potassium. Both these reactions involve the *reversal* of partial reactions of the pump enzyme. Consider Fig. 6.3 carefully: if the pathways involving sodium are reversible, then when no potassium is present to bind to the enzyme, sodium ions will be *exchanged* across the membrane (as Patricio Garrahan and Ian Glynn showed in the experiment depicted as the lower curve in Fig. 6.4). This reaction needs the presence of ATP and ADP (the latter to accept P from E_1PNa) but results in no overall splitting of the ATP. Similarly, potas-

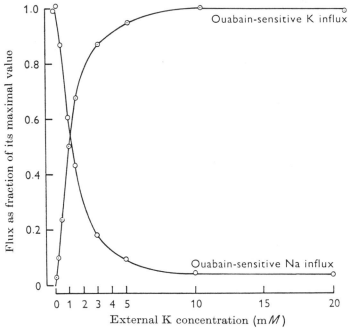

Fig. 6.4. Pump-mediated cation exchanges across red cell membrane. Fresh human red blood cells were incubated at 37°C in media containing labelled potassium (upper curve) or sodium (lower curve) and the ouabain-sensitive component of the cation influxes measured as a function of the external potassium concentration. The upper curve represents sodium–potassium exchange by the pump; the lower curve, sodium–sodium exchange. Taken, with kind permission, from P. J. Garrahan and I. M. Glynn (1967). *J. Physiol. (London)* **192,** 189–216.

(Continued)

sium ions (when little or no sodium is present) can be exchanged in a reaction that is speeded up by ATP and P_i but results in no hydrolysis. Part of the upper curve in Fig. 6.4 reflects this flux.

Some other properties of the system do not follow from Fig. 6.3 and, indeed, show that the scheme is an oversimplification. There is a slow, but definite, exchange of potassium ions in the absence of any phosphate or possibility of phosphorylation, and there is also a net transport of potassium ions in such conditions. Finally, there is a slow hydrolysis of ATP in the complete absence of potassium ions, accompanied by the transfer of sodium from intracellular to extracellular surface. This means that there is a route for the (slow) breakdown of E_2P in the absence of potassium ions (as can be seen, indeed, in Fig. 6.2). All these reactions are true "slippages" of the pump enzyme. (See Section 5.2.2 for "slippage" on co- and countertransporters.)

Just as we saw for the carrier (Section 4.2) and for the anti- and symporters (Sections 5.2.1 and 5.3.2), one can study the kinetics of the sodium and potassium transport on this primary transporter. One can measure a Michaelis constant K_m for each of the cationic substrates, sodium and potassium, and for the phosphate derivatives, ATP, ADP, and P_i, where binding is studied for each of the two forms E_1 and E_2. (Note that ATP, ADP, and P_i can bind to each form, E_1 and E_2, although their binding sites exist only at the intracellular face of the membrane.) The Na^+,K^+-ATPase is quite clearly asymmetric. The form E_1 has a high affinity for sodium and for ATP and ADP, but a low affinity for potassium and P_i. In contrast, E_2 has a high affinity for potassium and P_i, but a low affinity for sodium, ATP and ADP. The high affinity of E_1 for sodium and low affinity for potassium enable the pump to bind sodium in preference to potassium at the intracellular face (where the binding sites of E_1 are accessible) and this maximizes the *rate* of pumping of sodium out of the cell. Similarly, the high affinity of E_2 for potassium and low affinity for sodium enables the pump to bind potassium preferentially at the extracellular face and maximizes the rate of inward pumping of potassium.

How does the splitting of ATP drive the active transport of the ions? It does so as a result of the strict coupling that occurs between the conformational interchanges of E_1 and E_2, and the splitting of ATP. Coupling ensures that a complete cycle of ion transport and ATP splitting takes place only when (1) the sodium ions from the intracellular face become bound to E_1, (2) the conformation alters to E_2, (3) at the extracellular face potassium exchanges for the sodium ions, and finally (4) the conformation returns to E_1, bringing potassium to the intracellular face. Each time that a net breakdown of ATP and ADP and P_i occurs, the exchange of sodium and potassium has to occur. The ATP is a "high-energy" compound in that the equilibrium between ATP and ADP + P_i is well over on the side of the splitting of ATP at the usual levels of ATP, ADP, and P_i in the cell. Thus, a reaction leading to hydrolysis of ATP tends to occur. If this reaction occurs on the pump, the exchange of sodium with potassium will, under physiological conditions, be coupled with it, in the direction of sodium efflux and potassium influx (Fig. 6.3). But this overall reaction, like any chemical reaction, is, in principle, reversible. If the resources of ATP in the cell are low in comparison with the concentrations of ADP and P_i, the reaction of ATP hydrolysis might not be so favorable, energetically. If, in addition, extracellular levels of sodium are high and those of potassium intracellularly are high, a reversal of pumping will occur. Sodium will enter the cell and potassium will leave it, on the pump. The direction of travel around the scheme of Fig. 6.3 will be reversed and ATP *synthesis* will occur. Thus it is ATP splitting, driving the interconversion of E_1 and E_2, that brings about pumping. The change of affinity for the cations during this interconversion speeds pumping but does not drive it.

We can be somewhat more precise about *how* these changes of affinity for the cations are brought about. We have shown in Section 4.6.2, where we considered the asymmetry of carriers, that a transporter can exist in two conformational states that differ in free energy. The binding of a substrate will appear to be of low affinity when it binds to the form of lower free energy. This enables us to understand the affinity changes that the sodium–potassium pump undergoes. The change of affinity for potassium is brought about without any hydrolysis of the ATP. (Indeed, ADP is almost as good as ATP at doing this.) Adenosine triphosphate binds strongly to the E_1 form of the enzyme, poorly to the E_2 form. The mere *binding* of ATP stabilizes the E_1 form to which potassium ions are therefore *less* strongly bound. For sodium ions, Fig. 6.3 reveals that sodium binds to E_1 together with ATP, but to E_2 to the form that has bound P_i. Thus the two forms to which sodium binds have a different free energy, differing by the free energy of ATP hydrolysis, and E_2P is the form of low free energy. This form, therefore, has a low affinity for sodium ions. (Box 6.2 discusses the electrogenic nature of the Na^+,K^+-ATPase.)

6.1.2. Structure of the Sodium Pump

The Na^+,K^+-ATPase enzyme has been purified from many different types of cell membranes, often solubilized with detergents. The enzyme

Box 6.2. Electrogenicity of the Na^+, K^+-ATPase

Since the Na^+,K^+-ATPase pump enzyme extrudes three sodium ions for every two potassiums that it pumps inwards, there is a net efflux of one positive charge during each turn of the pump cycle. The pump is, therefore, electrogenic. The actual amount of current flow is given by the product of (1) the number of pumps in the cell membrane surface, (2) the rate of turnover of the pump (the number of cycles it performs per unit time), and (3) the charge on the electron (1.60×10^{-19} coulombs). For pumps from many cell membranes, the maximum velocity of pumping is some 175 ATP molecules split per second at 37°C. In the human red blood cell about 200 pumps are present in the 137-μm^2 surface area of the cell. This gives rise to a *maximum* current flow of $200 \times 175 \times 1.6 \times 10^{-19}$ or 5.6×10^{-15} A, or a current of 2.8×10^{-15} A when the pump is working at physiologic levels of ATP, potassium, and sodium at the inner face and of potassium and sodium at the extracellular face. The resistance of the red cell membrane is approximately 1.1×10^{11} Ω. Thus a voltage of -0.3 mV [volts (amperes $\times$ ohms)] will be established by the pump in physiological conditions. In comparison with the prevailing membrane potential of -9 mV for the red cell, this is a very small contribution to the overall potential. In general, such seems to be the case for the sodium pump in many cell membranes. In squid axon, for instance, the pump seems to contribute only some 5 mV for a total transmembrane potential of some 60 mV. Only in very specialized cells is the electrogenic contribution of the sodium pump significant. (Full details of such measurements can be found in the book "Electrogenic Transport" edited by M. Blaustein and M. Lieberman, Raven Press, New York, 1984).

can be reconstituted in an active form in artificial phospholipid vesicles across which transport of ions can be accurately measured and compared with the concomitant splitting of ATP. The purified enzyme has been shown to be composed of two polypeptide chains (termed, of course, α and β) some 112,000 and 34,000 Da in molecular mass. The genes coding for these chains have been cloned by Jerry Lingrel and co-workers in the United States, and by Shosaku Numa and colleagues in Japan.

A hydropathy plot of the sequence of amino acids predicted from the DNA sequence shows about eight regions (seven on the α chain and one on the β chain) that would appear to cross the membrane. This conclusion is consistent with chemical studies on the enzyme. One region of the protein can be shown directly to bind the ATP and is homologous in sequence to the nucleotide-binding region of certain soluble ATPases. A third region has been shown to be the site of phosphorylation of the enzyme. These data enable the following tentative model of the physical structure of the pump enzyme to be proposed [Fig. 6.5(a)], with its interconversion between the two conformations, E_1 and E_2, depicted in Fig. 6.5(b).

We sorely lack accurate information as to what part of the structure binds the cations and as to the path that the cations take as they cross the membrane through the pump enzyme. We shall see in Section 6.2 that a similar model, but much more detailed, has been proposed for the calcium-activated ATPase of the sarcoplasmic reticulum membrane of the muscle.

(a)

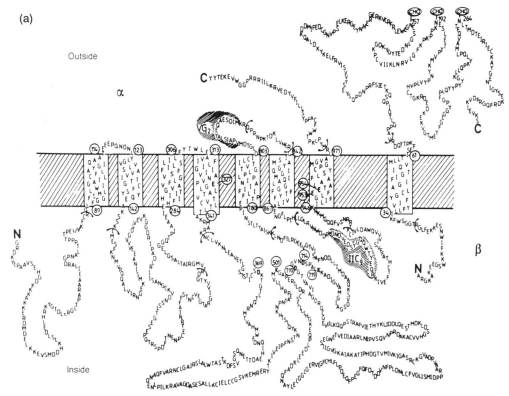

Fig. 6.5.   (*Figure continues*)

6.2. THE CALCIUM PUMP OF SARCOPLASMIC RETICULUM

The contraction and relaxation of muscle are controlled by the cytoplasmic concentration of free calcium ions. Calcium is stored within a cytoplasmic organelle, the sarcoplasmic reticulum (SR). In response to a signal transmitted to the tubular membrane system of the muscle across the nerve muscle end plate (see Section 3.2), calcium is rapidly released from the SR and binds to a protein, troponin, in an event that initiates contraction of the muscle (Fig. 6.6). Relaxation of the muscle follows when the cytoplasmic calcium concentration is lowered again, under the action of the calcium pump of the sarcoplasmic reticulum, a system studied originally by Wilhelm Hasselbach and Anne-Marie Weber. These researchers demonstrated the presence of an ATPase in the SR membrane that very effectively pumps calcium from the cytoplasm into the lumen of the SR vesicle. The SR membrane is packed tightly with molecules of this Ca²⁺-ATPase, which constitutes its major protein. This high concentration of the pump enzyme makes the SR a very handy object for obtaining

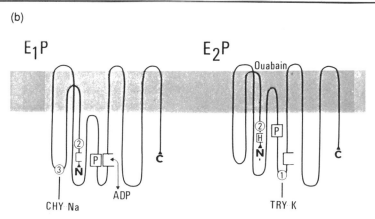

(b)

E_1P E_2P

CHY Na   ADP   TRY K

Fig. 6.5 (*Continued*). A model for the structure and the conformational change of the sodium–potassium pump. (a) Disposition of the α and β chains in the membrane, based on sequence data and studies with chemical reagents. The amino acids along the chains are numbered. Symbols N and C represent the amino and carboxyl-termini; hatched areas, antibody-binding regions; crescents, protease sites. Symbols CHO enclosed by a circle indicate sites of glycosylation in the β chain. Taken, with kind permission, from Y. A. Ovchinnikov (1987). *TIBS* **12**, 437. (b) Scheme for how the two conformations of the α-chain interconvert. The chains are depicted as situated within a lipid bilayer (the gray band). Symbols N and C represent the amino and carboxyl-termini, P the site of phosphorylation, and ADP the site of nucleotide binding. The primary sites of chymotryptic and tryptic digestion are shown. (Original figure, courtesy of Peter Jorgensen.)

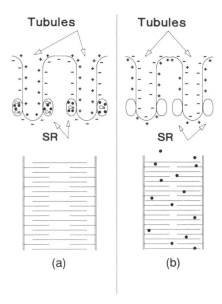

Fig. 6.6. Calcium controls the relaxation–contraction cycle of muscle. The upper half of the figure shows schemes for the tubular membrane system of muscle with attached sarcoplasmic reticulum (SR) vesicles (indicated by ellipses). The lower half of the figure shows the actomyosin bridge complexes of striated muscle. On the left (a) the muscle is relaxed with calcium (solid circles) sequestered within the SR. (b) The cell membrane has depolarized, and calcium has left the SR and become attached to the muscle proteins, bringing about the contraction step. Relaxation back to (a) occurs when the SR can once again pump the calcium ions back into itself on the ATP-driven pump.

pure preparations of a cation pump, ensuring that the calcium pump from the SR membrane has been an object of intense research over the years.

The initial study of the primary sequence of the protein led to the isolation of the gene that codes for the protein and to the determination of

Fig. 6.7. Possible structures of the SR calcium ATPase. (a) A speculative arrangement of the transmembrane helices and cytoplasmic regions of the calcium pump, as deduced from the hydropathy plot of the amino acid sequence, sequence homologies with other ATPases, and chemical evidence. The diagram shows six postulated "domains" of the structure. Three are thought to be located wholly in the cytoplasmic portion of the pump, the phosphorylation, nucleotide-binding, and "transduction" domains. The "stalk" connects these with the hydrophobic domain. In addition a "hinge" is postulated. The heavy, continuous lines show sequences that are well conserved between different cation pumps. (b) A speculative three-dimensional arrangement of the domains of (a). Both figures taken, with kind permission, from N. M. Green, W. R. Taylor, and D. H. MacLennan (1988) *in* "The Ion Pumps" (W. D. Stein, ed.), pp. 15–24. Alan R. Liss, New York.

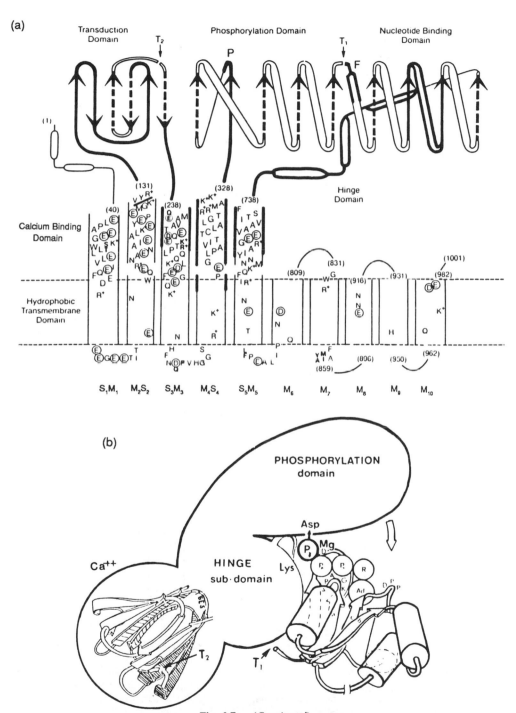

Fig. 6.7. (*Continued*)

the full sequence of the protein itself. Figure 6.7(a) is a representation of parts of this sequence, in a diagram that illustrates current attempts to fit the sequence to a tentative structural model of the enzyme. The protein probably spans the SR membrane ten times along α-helical hydrophobic segments. A part of the protein has been shown chemically to bind the ATP, and another region accepts the phosphate residue during the turn-over of the pump. These domains are marked on the diagram. A "hinge" domain links these regions to the membrane-spanning sequences, via a stalk that, it was originally believed (but see Box 6.3), contains calcium-binding areas. The portions of the sequence drawn with a heavy, black line are those regions that are highly conserved when the sequences of different vanadate-inhibitable pumps are compared. We will return to this point when we compare the detailed structure of the different ATPases in Section 6.5. The regions of the polypeptide backbone that may be con-cerned in binding of the calcium ions have been delineated by site-directed mutagenesis (see Box 6.3 and Fig. 6.8). A schematic view of the structure in three-dimensional form is shown in Fig. 6.7(b).

The calcium pump of the sarcoplasmic reticulum has been a favorite object for attempts at crystallization of pump protein, and the direct visu-alization of the structure of the pump by electron microscopy. Figure 6.9

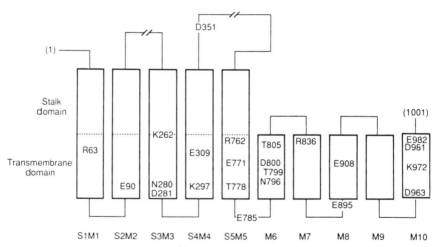

Fig. 6.8. Site-directed mutagenesis of the Ca²⁺-ATPase. Possible arrangement of mem-brane-spanning helices of the pump enzyme showing the sites at which mutations were introduced into the molecule. The text lists the six sites where the mutation altered the activity of the enzyme. The amino acid marked D351 (an aspartyl residue) was not mutated but is the site at which phosphorylation from ATP takes place. Taken, with kind permission, from D. M. Clarke *et al.* (1989). *Nature* **339**, 476–478.

Box 6.3. Site-Directed Mutagenesis for Delineation of the Calcium-Binding Groups of Ca^{2+}-ATPase

The cloning of the gene for the SR calcium ATPase has enabled Giuseppe Inesi and David MacLennan, with their colleagues, to use site-directed mutagenesis (see Box 3.3) to determine what amino acid side chains are present at the calcium-binding sites of the enzyme. Complementary DNA cloned from the rabbit fast-twitch muscle Ca^{2+}-ATPase gene was *transfected* into a line of mouse kidney cells grown in cell culture. (Transfection is the technique whereby foreign DNA is introduced into a recipient cell, often in the form of a calcium phosphate precipitate of the DNA.) These cells then expressed the ATPase in their microsomal membrane fraction at levels which could reach 50-fold that of the control, nontransfected cells. Mutations were introduced into the DNA at the positions marked with sequence numbers (and amino acid single-letter code names) in Fig. 6.8. In all, over 24 mutations were introduced (one at a time) into the enzyme structure. For each mutation, MacLennan and colleagues measured its effect on the ability of calcium to stimulate phosphorylation of the enzyme by ATP. (See Fig. 6.2 for phosphorylation of the sodium pump and Fig. 6.10, below, for phosphorylation of the calcium pump.)

Mutations at six positions in the sequence (at Glu-309, Glu-771, Asn-796, Thr-799, Asp-800, and Glu-908) prevented calcium from stimulating phosphorylation of the enzyme by ATP, but did not prevent phosphorylation of the enzyme by phosphate in the absence of calcium. This latter observation was an important *control* in that it showed that the site-directed mutations were not upsetting the whole structure of the enzyme and preventing it from undergoing any phosphorylation reaction. Notice that these six sites are in regions of the enzyme that the hydropathy plot suggests are in hydrophobic regions (although there is still some disagreement about the correctness of these identifications). Note, too, that it seems to be possible to arrange the hydrophobic helices in three dimensions so that the six residues identified by mutagenesis are contiguous. We cannot be sure that all six of these residues are indeed at the calcium-binding sites, but the evidence is suggestive and the technique capable of much further application. In another study, also using site-directed mutagenesis, MacLennan and colleagues showed that the so-called stalk region of the molecule (see Fig. 6.7) is *not* involved in calcium binding. Thirty of the carboxyl groups in this region of the molecule were replaced without significantly affecting the ability of calcium to be transported into SR vesicles.

shows electron micrographs prepared by Anthony Martonosi and his interpretations of these pictures for three different preparations of the calcium pump from the SR membrane.

Figure 6.9(a) shows an electron micrograph of crystals of the pump enzyme grown in the presence of calcium ions at pH 8. As these conditions stabilise the E_1 form of the enzyme, the electron micrograph shows, presumably, the protein in its E_1 form. The crystals form within the SR membrane itself, when this is held for several weeks under refrigeration, and the formation of the crystals is promoted if the transmembrane potential is held inside negative. The size of the unit cell of the crystal is consistent with monomers of the pump enzyme as the structural unit of the crystal. In complete contrast is the picture of the crystals that form from SR membranes when they are stored in the *absence* of calcium ions, but in the presence of vanadate ions [see Fig. 6.9(b)]. These conditions

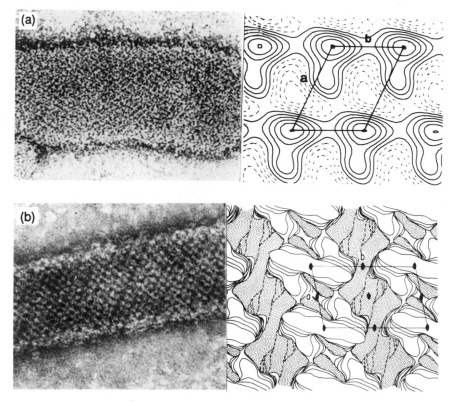

Fig. 6.9.   (*Figure continues*)

favor the E_2 form of the enzyme, and the formation of these crystals is promoted by holding the transmembrane potential inside positive. The size of the unit cell of the crystal is in this case consistent with there being two molecules of the pump enzyme needed to form the unit cell. The crystals are disrupted when calcium ions are added and leach into the crystals. Finally, Fig. 6.9(c) shows the structure of the lamellar sheets that form when detergent-solubilized pump enzyme is stored under refrigeration for several weeks in the presence of high concentrations of calcium. The interpretation of the electron micrograph shows ATPase headgroups extending out from the lipid–detergent layer, the "filling" of the "sandwiches" that compose this three-dimensional array.

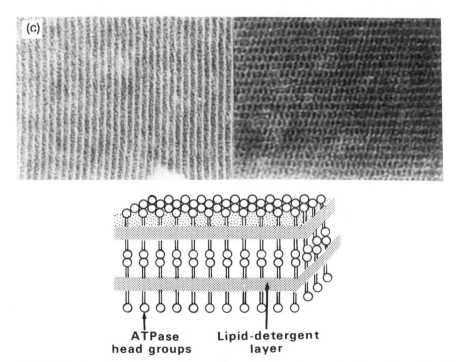

Fig. 6.9 (*Continued*).   Electron micrographs and deduced reconstructions of crystals of the SR Ca^{2+}-ATPase. (a) Micrograph and the deduced density projection map of two-dimensional crystals of the E_1 form of the Ca^{2+}-ATPase. (b) As (a) but for the E_2 form. (c) Three-dimensional crystals in detergent-solubilized SR Ca^{2+}-ATPase. Top left shows a thin section, top right a negatively stained preparation and the bottom drawing is a schematic representation of a crystalline lamella. All three figures taken, with kind permission, from Martonosi *et al.* (1988).

What can one learn from these pictures? First, they offer direct evidence that the pump enzyme can exist in more than one conformation. The two forms shown in Fig. 6.9(a) and (b) are almost certainly E_1 and E_2, respectively. That the E_2 crystals disintegrate when calcium is added, favoring formation of the E_1 form, is again strong visual evidence that pump action is linked to a change in the conformation of the enzyme. The electron micrographs confirm the size of the pump enzyme, since the size of the unit cell is consistent with the sizes determined by other, more indirect methods. The pictures are not as yet detailed enough to justify attempts to fit the protein sequence data to the overall protein structure, but the progress made [especially with the three-dimensional crystals of Fig. 6.9(c)] gives hope that such detailed fitting will be possible within a few years' time.

The calcium-pump enzyme is very similar in its properties to the Na^+,K^+-ATPase that we discussed in Section 6.1. As we have just seen, it can exist in two major conformations, and just as in the Na^+,K^+-ATPase, the two forms give different fluorescent signals. Thus rates of interconver-

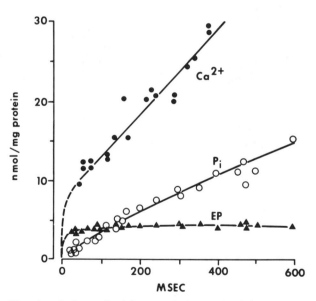

Fig. 6.10. Phosphorylation and calcium uptake by the SR Ca^{2+}-ATPase. Formation of the phosphoenzyme (filled triangles), phosphate liberation (open circles), and calcium uptake (filled circles) were measured with SR membranes at 25°C. At zero time 100 μM ATP was added in the presence of KCl and calcium ions. From the ratio of the slopes of the upper two curves, the stoichiometry of phosphate liberated and Ca^{2+} can be seen to be 1:2. Taken, with kind permission, from G. Inesi *et al.* (1982). *Ann. N. Y. Acad. Sci.* **402,** 515–532.

Box 6.4. Stoichiometry of Ion Pumping

We saw in Sections 5.2.4 and 5.3.4 how important, physiologically, the stoichiometry of coupling is in the case of counter- and cotransport. Stoichiometric considerations are equally important in the case of the primary transporters, where the concentration ratio of the pumped ion will be determined by the *stoichiometry* of the pump for its cation substrate(s). Since, however, we have to take into account the change in chemical potential that occurs during the progress of the chemical reaction of ATP hydrolysis, the concept is a little more difficult. But the problem is still easy to handle, and we proceed along the following lines:

We know that for the chemical reaction ATP→ADP + P_i we can write the equilibrium constant for the reaction $K_{ATP \rightarrow ADP + P_i}$ in terms of the concentrations of ATP, ADP, and P_i at equilibrium as

$$K_{ATP \rightarrow ADP + P_i} = [ADP][P_i]/[ATP] \qquad (B6.4.1)$$

If, instead, we consider ATP hydrolysis linked to the transport of a number m of metal ions M from side 1 of the membrane (where the metal is at concentration M_1) to side 2 of the membrane (where it is at concentration M_2), the reaction can be written as ATP + mM_1→ADP + P_i + mM_2. For this reaction, we can write the *equilibrium condition* as

$$K_{ATP \rightarrow ADP + P_i} = [ADP][P_i]M_2^m/[ATP]M_1^m \qquad (B6.4.2)$$

We can do this because the metal is in the same chemical state on the two sides of the membrane. There is thus no change in its standard-state chemical potential (see Section 2.1) and we do not have to include a term for the "equilibrium constant" for any chemical reaction of the metal ion. We can now rewrite Eq. (B6.4.2) so as to emphasize what determines the steady-state distribution of M:

$$(M_1/M_2)^m = [ADP][P_i]/[ATP]K_{ATP \rightarrow ADP + P_i} \qquad (B6.4.3)$$

Clearly, from Eq. (B6.4.3) if the concentration of ATP is high, compared with that of ADP and P_i *divided by the equilibrium constant* $K_{ATP \rightarrow ADP + P_i}$, the concentration of M_2 will be high compared with that of M_1, i.e., M will be pumped from side 1 (the cytoplasmic face of the cell or SR membrane) to side 2. A given ratio of ATP, ADP, P_i, and the equilibrium constant $K_{ATP \rightarrow ADP + P_i}$, will create a steeper concentration ratio of the metal ion M, if the stoichiometric coefficient is less.

(Continued)

Take, as an example, values of ATP, ADP, and P_i concentrations of 5 mM, 50 μM, and 5 mM, respectively, as representative of those that occur in many living cells. Take $K_{ATP \to ADP+P_i}$ equal to 10^6, again a reasonable approximation to its value at 37°C. Substituting in Eq. (B6.4.3), we see that for $m = 1$, i.e., a stoichiometry of unity, we would expect to find a metal ion concentration ratio of $(10^6 \times 10^3)/(0.05)$ at equilibrium, i.e., of 2×10^{10}. Such enormous concentration ratios are never found in living cells! With the stoichiometric coefficient of 2, as found for the calcium pump in Fig. 6.10, $m = 2$ and, therefore, the concentration ratio at equilibrium will be the square root of the number just calculated or 1.4×10^5. This is not too far from the value often encountered for the SR vesicle, where cytoplasmic calcium concentrations can approach 30 nM and intravesicular concentrations, 1 mM, a ratio of 3×10^4.

Consider, now, the sodium–potassium pump discussed in Section 6.1. Here the stoichiometry of pumping of sodium is 3 and that of potassium, 2, for each ATP split. We need to rewrite Eq. (B6.4.3) in terms of two types of metal ions, M and N, each with its own stoichiometry m and n. The reader can check that with prevailing cation ratios of about sevenfold for sodium (inside low) and 20-fold for potassium (inside high), prevailing levels of ATP, ADP, and P_i are such as to be consistent with the predictions of the stoichiometry of the pump.

The reader will realize that the treatment given above is an oversimplification since we have omitted any consideration of the electrical potential. (We saw in Box 6.2 that the sodium pump, at least, is indeed electrogenic.) Insofar as an ion pump is electrogenic, i.e., if it transfers n units of charge across the membrane during a complete cycle of pumping, we have to modify Eq. (B6.4.3) to take the transmembrane potential ψ into account. The complete equation is

$$\ln[(M_1/M_2)^m] + (nF/RT)\Delta\psi = \ln([ADP][P_i]/[ATP]K_{ATP \to ADP+P_i})$$

sion of the two conformations can be measured. The absolute rates are rather similar to those of the Na$^+$,K$^+$-ATPase. Calcium binds to the E_1 form with high affinity, as this is the form in which calcium-binding sites face the cytoplasmic surface of the SR vesicle. The E_2 form binds calcium with far lower affinity. ATP binds to the E_1 form with high affinity, P_i with low affinity, while ATP binds to E_2 with a low affinity, P_i with high. The enzyme is phosphorylated when calcium binds to E_1, following addition of ATP (see the curve labeled EP in Fig. 6.10). There is, however, no equivalent of the potassium ion to activate the dephosphorylation of the pump enzyme, a step that occurs spontaneously at a rapid rate (the curve labeled P_i in Fig. 6.10). Figure 6.10 also shows the rate of uptake of calcium ions into the SR vesicles (the curve marked Ca^{2+}). Note that the slope of this curve is twice that of the rate of formation of P_i, i.e., of the rate of splitting of ATP. It follows that the stoichiometry of calcium ions pumped to ATP molecules split is 2:1. There is more to be learned from a careful study of this important experiment. Consider the intercept to zero time of the curves for Ca^{2+} and P_i uptake. Almost twice as much calcium as phosphate is bound very rapidly to the enzyme. This is calcium that is not taken up *within* the vesicles, but is rather *bound* to the enzyme itself. This is "occluded" calcium (see Section 6.1.1 for a similar finding for the sodium pump). Clearly, two ions of calcium are occluded per molecule of the calcium pump. (Interestingly, detailed studies by Incsi of this occlusion show that of the two calcium ions that are bound to the pump at the cytoplasmic face, the one that is *bound* first is the first to be *released* at the extracellular surface after the protein's conformation change has occurred. Thus the calcium ions seem to "glide" through the protein interior!)

Vanadate, as we have seen, binds the calcium pump of SR membranes in place of phosphate to form a stable complex, inhibiting further action of the enzyme. There is only one polypeptide chain in the enzyme, in contrast to the Na^+,K^+-ATPase. The calcium pump, as we saw, occludes the calcium as the ion is transferred across the membrane, just as the Na^+,K^+-ATPase binds *its* substrates, sodium and potassium. The strong analogy between the calcium pump and the sodium–potassium pump points to a common mechanism of action. Of course, both pumps are driven by the act of ATP splitting, and the reaction of splitting is strongly favored at the prevailing concentrations of ATP, ADP, and P_i in the cell. Hence, the cycling of the calcium pump in the direction of removal of calcium from the cytoplasm is favored. This pump is, again, reversible, and ATP synthesis from ADP and P_i will occur when levels of calcium are such as to favor this direction of cycling. It is the strict coupling of calcium transport and ATP splitting that drives transport. The fact that the calcium pump changes its affinity from high to low as it alters conformation from E_1 (with binding sites facing the cytoplasm) to E_2 (with binding sites facing the lumen of the vesicle) speeds up the transport of calcium but is not the force that drives it.

6.3. THE CALCIUM PUMP OF THE
PLASMA MEMBRANE

The sodium–calcium antiporter, discussed in detail in Section 5.2.5, is only one of the systems by which the cytoplasmic concentration of calcium ions in the cell is kept low. The other method is the harnessing of primary active transport by the calcium-activated ATPase of the plasma membrane, a system first identified by Hans Schatzmann. Such a primary transporter of calcium ions is present in the plasma membrane of most animal cells. Two well-studied systems are those of red blood cells and the nerve membrane. The system in the human red blood cell is very effective. With cells stored in their own plasma, at a free calcium concentration of 1.3 mM, the intracellular free calcium can be as low as 26 nM, a 50,000-fold concentration ratio. The transmembrane potential in these red cells is about 10 mV, so that efflux of a divalent cation is hindered by the inside negative potential by a factor of 2.25-fold (59 mV is equivalent to a 100-fold concentration ratio for a divalent cation; see Section 2.2). All in all, calcium ions are pumped out against an equivalent gradient of over 10^5! (Recall from Box 6.4 that splitting of an ATP molecule can cope with this equivalent concentration ratio using a stoichiometry of two. This is, indeed, the measured stoichiometry of the calcium-pump enzyme of the plasma membrane.)

In its properties, the calcium pump of the plasma membrane strongly resembles the Na^+,K^+-ATPase and the SR Ca^{2+}-ATPase that we discussed in the previous sections. This system, too, is inhibited by vanadate, forms an isolatable phosphoenzyme, has a high affinity for its substrate at the cytoplasmic face of the membrane, a low affinity at the extracellular surface. Like the calcium pump of the SR, it is not inhibited by ouabain. It is composed of a single polypeptide chain. What is quite remarkable about the Ca^{2+}-ATPase of the plasma membrane is that its rate of action can be precisely *regulated* by a cytoplasmic factor, a property not shared by the sodium pump that we discussed previously. As we discussed in Section 5.2.5, the free calcium ion concentration of the cell is a major target for the transduction of extracellular signals received by the cell. Wiping out the memory of these signals requires that cytoplasmic levels of calcium be restored rapidly and effectively. The calcium pump plays its part in this task, acting in a regulated fashion. As calcium levels rise, its rate of pumping speeds up and does so far more rapidly than would be the case were it to be activated solely by the direct binding of calcium.

Consider Fig. 6.11(a), which depicts the activity of the calcium pump as a function of the concentration of ATP, for the isolated pump enzyme [lower curve, Fig. 6.11(a)] and for the same preparation but in the presence of a cytoplasmic protein named calmodulin [upper curve, Fig. 6.11(a)]. Clearly, calmodulin greatly increases the rate of calcium pumping at all ATP concentrations, and is even more effective at the higher concentrations. Calmodulin is a water-soluble cytoplasmic protein, with a molecular mass of 16,700 Da. A single molecule of calmodulin binds four calcium ions with very high affinity. Calmodulin is a member of a large family of calcium-binding proteins, all sharing the property of possessing a particular concatenation of amino acids at their calcium-binding sites, the so-called calcium hand. Calmodulin binds to the plasma membrane calcium pump, but only when the calmodulin has bound to four calcium ions. Binding of calmodulin to the pump modifies its activity in the manner shown in Fig. 6.11(a). Thus when concentrations of calcium in the cytoplasm rise, following some extracellular signal, some of these incoming calcium ions become bound to calmodulin, its own binding to the Ca^{2+}-ATPase takes place and the pump enzyme acts more rapidly, increasing the rate at which calcium is expelled from the cell. As the level of calcium drops again, below the levels at which all four calcium ions can bind to calmodulin, calmodulin dissociates from the pump enzyme and pumping slows down again. Thus it is the affinity of calmodulin for calcium ions that helps to regulate the rate of calcium pumping and hence the concentration of cytoplasmic calcium. Phosphorylation of calmodulin can control the pump-calmodulin interaction.

How does calmodulin have this dramatic effect on the rate of calcium pumping? Figure 6.11(b) (upper half) shows how the activity of the pump enzyme is affected by the concentration of *calcium* ions, in the presence and absence of calmodulin. Again, there is a marked stimulation by calmodulin. These data are for the isolated plasma membrane Ca^{2+}-ATPase, reconstituted into phospholipid vesicles made from the neutral phospholipid, phosphatidylcholine. If the same experiment is performed using vesicles made with the phospholipid phosphatidylserine, a negatively charged molecule, no effect of calmodulin is found; the enzyme is fully active even in the absence of the activator [lower half of Fig. 6.11(b)]. A fully active enzyme is also obtained when the isolated pump enzyme is digested with trypsin, a treatment which splits off a 35,000-molecular-weight polypeptide from the enzyme. It would seem that this polypeptide somehow partially blocks access of ATP and Ca^{2+} to its active sites on the pump enzyme. When the peptide is removed by tryptic digestion, or when the enzyme is in a region of local negative charge, or when calmodulin is bound to it, the effect of the blocking peptide is removed, and full activity is obtained.

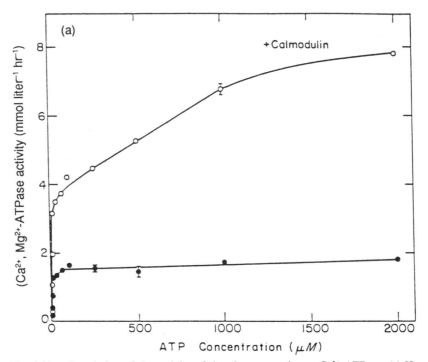

Fig. 6.11. Regulation of the activity of the plasma membrane Ca^{2+}-ATPase. (a) How calmodulin affects the affinity of the pump for ATP and also its maximum velocity. The enzyme, from human red blood cell membranes, was assayed at 37°C in the presence of 100 mM NaCl, 10 mM KCl, ouabain, and 50 μM calcium. The upper curve shows data when 150 μM calmodulin was added; the lower curve, in the absence of calmodulin. Taken, with kind permission, from S. Muallem and S. J. D. Karlish (1980). *Biochim. Biophys. Acta* **597**, 631–636. (b) How calmodulin and the phospholipid content affects the affinity of the pump for Ca^{2+}. The upper panel shows data where the membrane vesicles were reconstituted with phosphatidylcholine; the lower panel, with phosphatidylserine. The circles are in the absence of calmodulin; the triangles, in its presence. Data at 37°C in 120 mM KCl. Taken, with kind permission, from K. Stieger and S. Lauterbacher (1981). *Biochim. Biophys. Acta* **641**, 270–275. (*Figure continues*)

6.4. THE H$^+$,K$^+$-ATPase OF GASTRIC MUCOSA: THE PROTON PUMP OF THE STOMACH

The lining of the stomach secretes into the lumen of the stomach a highly acid secretion, which can reach as high as 0.16 M HCl. Specialized cells in this lining, the parietal cells, have the function of providing this secretion. Proton pumps are present within the cell in the form of intracellular vesicles. As a result of a hormonal signal, these vesicles fuse with the plasma membrane of the cell, the pumps being inserted into the mem-

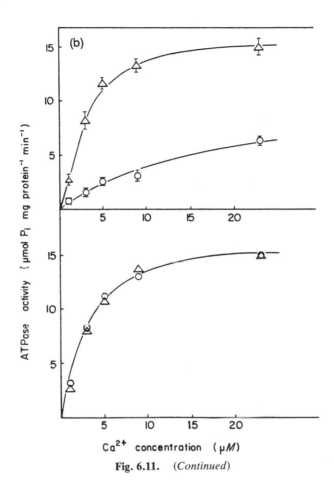

Fig. 6.11.   (*Continued*)

brane. The cells can develop a pH difference, inside low, of about 6.6 units of pH. This is equivalent to a concentration ratio of 4×10^6!

How is this achieved? Data obtained in the laboratory of George Sachs make this clear. Figure 6.12(a) shows the phosphorylation by ATP of a membrane preparation obtained from the acid-secreting vesicles of gastric mucosa. The filled symbols in Fig. 6.12(a) show phosphorylation when the extravesicular pH is 5.5, while within the vesicle the pH is 5.5, 6.7, or 8.0. At all internal pH values, ATP is split rapidly to phosphorylate the membrane proteins. In contrast, the empty symbols show data for similar vesicles having the same values of internal pH, but an extravesicular pH of 8.0. Clearly, phosphorylation of the membranes is achieved rapidly only when there is an acid pH externally, i.e., when there are sufficient protons at the extravesicular surface. The intravesicular pH does not

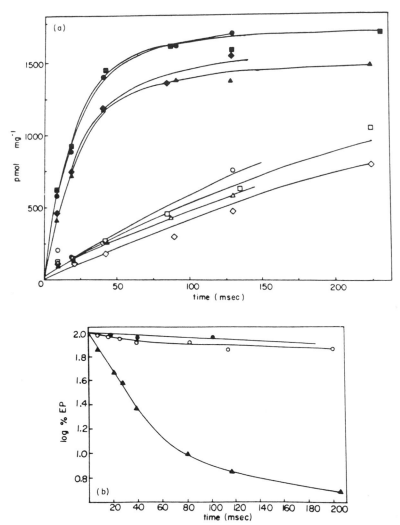

Fig. 6.12. Phosphorylation and dephosphorylation of the gastric H^+,K^+ ATPase. (a) Phosphorylation of the enzyme, prepared from pig stomach at 22°C. The filled symbols represent vesicles prepared with an extravesicular pH of 5.5 and various intracellular pH values. The empty symbols denote vesicles prepared with an extravesicular pH of 8.5 and various intracellular pH values. (b) Dephosphorylation. As above, but phosphorylated membrane vesicles were allowed to dephosphorylate by adding CDTA in order to complex magnesium. The triangles show vesicles to which potassium chloride had been added and also gramicidin, an ionophore, to allow the potassium access to what is the extracellular surface of the membranes. In contrast, the controls (circles) had no potassium added, or potassium added with no gramicidin, so that the potassium could not reach the extracellular surface of the membranes. Taken, with permission, from Stewart *et al.* (1981). *J. Biol. Chem.* **256,** 2682–2690.

affect the rate of phosphorylation. In contrast, Fig. 6.12(b) shows that the rate of *dephosphorylation* is dependent on the concentration of potassium ions and that these potassium ions have to be present inside the vesicles, carried there by the ionophore gramicidin (see Section 2.1). Thus, overall, a hydrolysis of ATP is activated by protons at the cytoplasmic face and potassium ions at the lumen face of the vesicle. The pump is an H^+,K^+-ATPase and closely homologous with the Na^+,K^+-ATPase.

The H^+,K^+-ATPase from gastric mucosa has been highly purified and shown to consist of two polypeptide chains, and the gene coding for the longest chain—that binding to and phosphorylated by ATP—has been cloned. The amino acid sequence of this chain is more than 60% identical to the sequence of the long (α) chain of the Na^+,K^+-ATPase. It is clear that two protons take the place of the three sodium ions pumped by the Na^+,K^+-ATPase, while the potassium ion has the same function in both systems. The stoichiometry of pumping is thus two H^+ and two K^+ transported for every ATP that is split. Researchers agree that the proton and sodium pumps must have strong evolutionary links, the Na^+,K^+-ATPase being almost certainly the progenitor of this proton pump. (Indeed, as Rhoda Blostein has shown, in the absence of sodium ions the sodium pump can be activated by protons, so that it will split ATP with only protons at the inner face of the membrane, and potassium ions at the outer face. Here, the sodium pump is acting as an H^+,K^+-ATPase!) Why should the gastric proton pump move two protons rather than three? It must be stoichiometric considerations that have forced the evolution of this characteristic. A pH gradient of 6.6 is above the steepest that an ATP-driven pump can achieve if the stoichiometry is two ions pumped (see Box 6.4 for a discussion of stoichiometry). Three ions cannot be pumped simultaneously against a gradient higher than 10^4-fold. Indeed, it even seems likely that the stoichiometry of the H^+,K^+-ATPase is *variable*, so that at the very highest gradients, one rather than two protons are transported for each ATP hydrolyzed.

Box 6.5. Multidrug Resistance and ATP-Driven Drug Pumping

From the earliest days of cancer chemotherapy it has been known that cancer cells quickly become resistant to the action of drugs that previously were effective against them. This phenomenon of resistance remains one of the major obstacles to the development of useful anti-

(Continued)

cancer drugs. In the last few years it has become clear that at least one route by which a clone of cancer cells becomes resistant to chemotherapy is by its acquisition of an amplified ATP-driven system that pumps the drug out of the cell. It is clear, too, that such systems are active against a range of structurally unrelated drugs (such as colchicine, doxorubicin or adriamycin, actinomycin D, and vinblastine). They are multidrug resistance (mdr) systems.

A gene (mdr) has been identified and cloned, and its cDNA has been shown to convey drug resistance to drug-sensitive cells when it is transfected into them. The gene codes for a polypeptide of molecular mass 170 kDa (and is thus often referred to as "P-170" but also as the P-glycoprotein) with an amino acid sequence suggesting the presence of 12 membrane-spanning regions. Interestingly, the polypeptide seems to be composed of two repeated halves with 43% homology between them, suggesting its evolution from a single half-structure by gene duplication. Each half is strikingly homologous with the sequences of certain bacterial transport proteins and contains a 230-amino-acid-long sequence that shows 47% identity with what is thought to be the domain concerned in ATP-binding in these transporters.

Membranes prepared from drug-resistant cells form vesicles that can accumulate the anticancer drug vinblastine in the presence of ATP. (Presumably the vesicles are *inside–out*, which is why they accumulate the drug rather than extrude it as do the intact cells). The accumulation of the drug is inhibited by vanadate, which, as we have seen, inhibits the E_1E_2-ATPases. Such inhibition is noncompetitive but among themselves the different drugs compete with one another.

It has been shown that the *mdr* gene is present in multiple copies in drug-resistant strains of cells and that the degree of resistance parallels the numbers of copies of the *mdr* gene. The gene can be present in minute chromosomes in these cells, a feature often associated with the development of drug resistance. The gene is, of course, expressed in normal cells in the body, but only in certain specific locations: colon, small intestine, kidney, liver, and adrenal tissue. The first four of these are tissues that might be expected to encounter toxic compounds from the environment and the localization in those tissues of transport systems that keep such substances out of the cell (and hence out of the body) is very probably of physiologic importance. It would seem that these physiologically useful systems are abnormally expressed in cancer cells giving them a survival advantage to the detriment of the survival of the whole animal.

In all respects, the H^+,K^+-ATPase is a close relative of the other E_1E_2-ATPases. Figure 6.12 shows that it acts through a phosphorylated intermediate. It is inhibited by vanadate, exists in two conformations that interchange during transport, and has strong sequence homologies with the other E_1E_2-ATPases. In Section 6.5 we discuss these homologies in a little more detail.

6.5.   THE E_1E_2-ATPases COMPARED

We have emphasized the strong relationship between the functional properties of the E_1E_2-ATPases, including their possession of a phos-

(a)

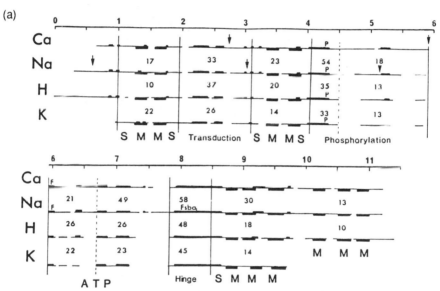

Fig. 6.13.   Sequence comparisons of the E_1E_2-ATPases. (a) Alignment of the sequences of four cation pumps. Symbols Ca, Na, H, and K represent the sequences of the fast skeletal muscle Ca pump, the sheep Na pump, the proton pump of yeast, and the K pump of *E. coli*, respectively. Conserved sequences are shown as bars above the line; hydrophobics, as bars below the line. ATP, nucleotide-binding site; S, the so-called stalk sequences; M, the putative transmembrane helical segments. The numbers above the lines are the identity percentages between the individual domains above and below the number. Taken, with kind permission, from N. M. Green, W. R. Taylor, and D. H. MacLennan (1988) *in* "The Ion Pumps" (W. D. Stein, ed.), pp. 15–24. Alan R. Liss, New York. (b) Hydropathy plots for five cation pumps. As in (a) except that "H, K" represents the plot for the α chain of the gastric proton pump. Here, the numbers on the right represent the overall sequence homology with the Na^+, K^+-ATPase. Taken, with kind permission, from Jorgensen and Andersen (1988). (c) A possible evolutionary tree for the radiation of the cation pumps. The upper section (α, etc.) depict the different sodium pumps. The percentages are the degrees of homology relative to the α subunit of the human Na^+, K^+-ATPase. Taken, with kind permission, from P. L. Jorgensen (1988) *in* "The Ion Pumps" (W. D. Stein, ed.), pp. 15–24. Alan R. Liss, New York. (*Figure continues*)

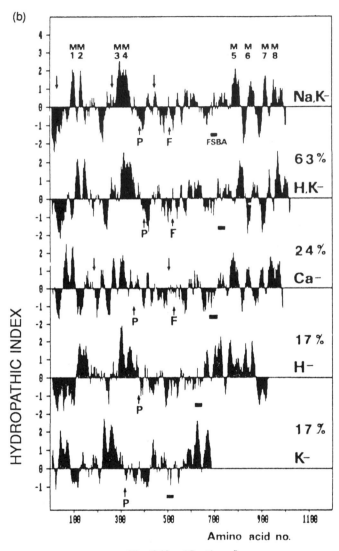

Fig. 6.13   (*Continued*)

phorylated intermediate, their inhibition by vanadate, and their existence in two inconverting conformations. Their amino acid sequences are also very closely related as Fig. 6.13(a) and (b) makes clear. The legend to Fig. 6.13 should be read carefully and includes data on two other E_1E_2 ATPases that we have not considered, a proton pump from yeast (labeled

(c)

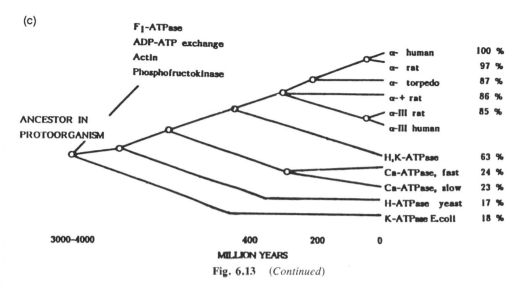

Fig. 6.13   (*Continued*)

H in Fig. 6.13), and a potassium pump from the bacterium *E. coli* (labeled K in Fig. 6.13).

It can be seen that the major regions of function are well conserved in all these proteins, although they come from such widely divergent organisms. It would be most unreasonable to suppose that such closely related proteins would not be related in an evolutionary sequence, perhaps the one proposed by Peter Jorgensen and depicted as the evolutionary tree of Fig. 6.13(c). The hypothesis that they are related would suggest that these cation pumps of such diverse and specialized function, pertaining to systems involved in such wholly "multicellular" actions such as muscular contraction, have their origin in the pumps of the unicellular and even prokaryotic organisms. One must expect that it is the common features of the pumps—those concerned with ATP binding, with phosphorylation and with the transduction of phosphorylation events into the signalling of the conformation changes—that are most highly conserved in the sequence. In contrast, it is in the regions that differ markedly in sequence that we might expect to find the regions of *specificity*, where the various cations, effectors and regulators are bound.

6.6. THE F_0F_1-ATPases

There is another great family of cation-activated ATPases. This is the family of proton-activated ATPases, known as the F_0F_1-ATPases, or

"F-ATPases." They are found in the membranes of the cell organelles, the mitochondria and chloroplasts, in bacterial membranes, and, as we shall see, a related pump is found in the membranes of the acidic vesicles in eukaryotic cells. The E_1E_2-ATPases of the previous sections of this chapter have, as their usual role, the *hydrolysis* of metabolically produced ATP, coupled to the pumping of the cations. In contrast, the F_0F_1-ATPases of the present section usually act to *synthesize* ATP, where synthesis is coupled to the movement of protons (or in some specialized systems, sodium ions) down their electrochemical gradients. The proton gradients are established by coupling of proton movements to the passage of electrons down the electron chain of the oxidation–reduction reactions of the cell. The F_0F_1-ATPases are often termed "ATP synthases," an appellation that emphasizes their special role in the cell. They certainly *can* act as ATPases and proton pumps, however, since the reactions that they undergo, like all other chemical reactions in the organism, are reversible. A major distinction between F_0F_1-ATPases and E_1E_2-ATPases is that only for the latter enzymes are phosphorylated intermediates formed during ATP hydrolysis. Since no phospho intermediates are formed, vanadate, too, does not combine with the F_0F_1 ATPases and they are therefore not inhibited by it. There seems to be little that links the two families of ATPases, either in an evolutionary sense or in the details of their structure and mechanism of action.

6.6.1. Structure of the F_0F_1-ATPases

The F_0F_1-ATPases have a far more complex structure than do the E_1E_2-ATPases. We saw that the latter are composed of one, two or at most three polypeptide chains. The F_0F_1-ATPases are built of no less than 20 chains! The whole ATPase is composed of two readily separable portions, F_0 and F_1. The F_0 part, which is an intrinsic membrane protein, is composed of a few polypeptides, I, II, III, and IV (Fig. 6.15), combining in a stoichiometry that has yet to be established. Subunit III is a low- (8000-Da) molecular-mass, membrane-spanning polypeptide of which about 12 copies are present in the functioning ATPase. The F_1 portion is formed of at least 10 polypeptides, of five different types, the whole F_1 comprising three copies of the α chain, three of the β chain and one or two of the γ, δ, and ε subunits; the structure of F_1 is written as $\alpha_3\,\beta_3\gamma\delta\varepsilon$. The β-chain has been shown to bear the ATP-binding sites, each β subunit being capable of binding a pair of nucleotide molecules. The genes coding for all of the five types of polypeptide have been cloned and sequenced. The α and β chains share a good deal of sequences in common, and sequences from very widely different organisms show much homology.

(a)

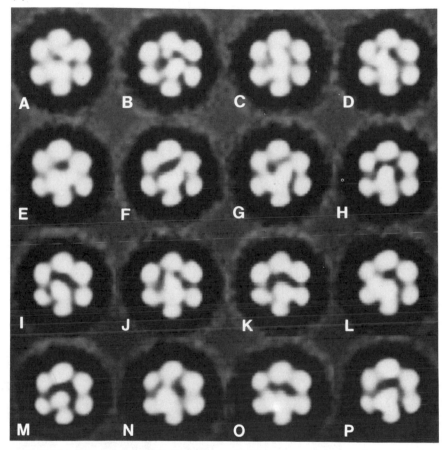

Fig. 6.14. Structure of the F$_0$F$_1$-ATPases. (a) Molecular projections of the chloroplast F$_1$-ATPase, as determined by computer-aided classification and aligning of 3300 electron-microscopic images. The subunit structure is clearly seen. (b) Three-dimensional model of the F$_1$-ATPase in side (above) and top (below) view. Dimensions refer to the beef heart system. (c) Electron micrographs in side view of the F$_0$F$_1$-ATPases from chloroplasts and mitochondria. Figures (a) and (c) taken, with kind permission, from E. Boekema, M. van Heel, and P. Gräber (1988) *in* "The Ion Pumps" (W. D. Stein, ed.), pp. 75–80. Alan R. Liss, New York. Figure (b) taken, with kind permission, from G. Schäfer *et al.* (1988) *in* "The Ion Pumps" (W. D. Stein, ed.), pp. 57–66. Alan R. Liss, New York. (*Figure continues*)

The F$_1$ portion is a water-soluble protein and functions as an ATPase. All are agreed that the transmembranal F$_0$ portion provides the route by which protons flow across the membrane; this flow through F$_0$ is coupled to the ATP synthesis performed by F$_1$. The coupling of proton flow and

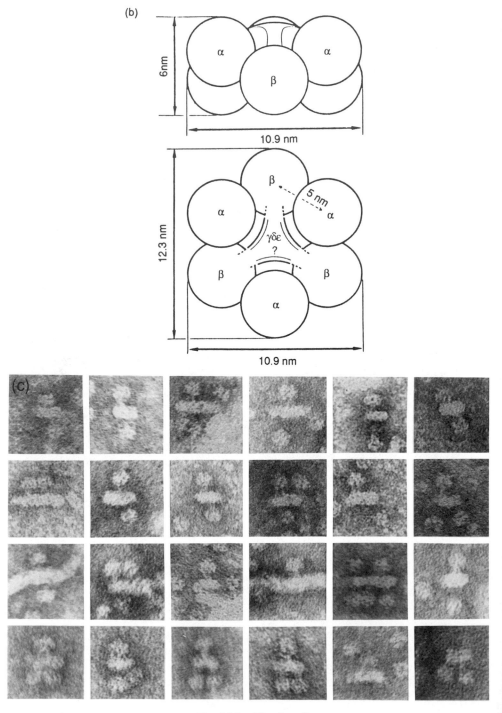

Fig. 6.14 (*Continued*)

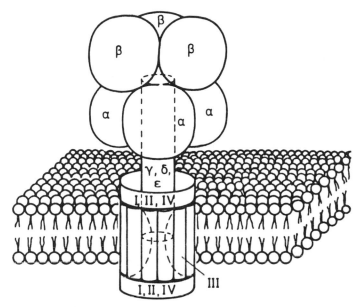

Fig. 6.15. A model of the $F_O F_1$-ATPase. Hypothetical scheme of the arrangement of the subunits in the chloroplast $F_O F_1$-ATPase. The cylinder, embedded in the lipid bilayer, is the F_O portion. Directly above it is the putative stalk, and above that again, the F_1 portion, the ATPase. (Taken, with kind permission, from P. Gräber *et al.* (1988) in "The Ion Pumps" (W. D. Stein, ed.), pp. 67–74. Alan R. Liss, New York.

ATP synthesis (or hydrolysis) can be broken, i.e., "uncoupled," by a number of treatments. In the most obvious of these, the F_O and F_1 portions of the whole enzyme are physically separated. For instance, as Ephraim Racker showed, when the intact membranes are treated with low-ionic-strength buffers, this breaks the F_O/F_1 connection leading to solubilization of F_1, while F_O remains bound to the membrane. Then F_1 acts as an ATPase and, since there is no membrane present, no flow of protons takes place across F_1. The chemical reagent dicyclohexylcarbodiimide (DCCD) inhibits proton flow in both F_O and the intact $F_O F_1$-ATPase, but not ATP synthesis (or hydrolysis) by the isolated F_1 portion. DCCD binds to F_O at a site on the chain III.

Electron micrographs of F_1 show that this can aggregate as an hexagonal array of subunits [Fig. 6.14(a)]. Detailed analysis of such electron micrographs, including side-on views of the protein, shows it to be composed of a double array of trimers, the α- and β-chain triplets being arranged in two separate layers [Fig. 6.14(b)]. Attached to this is a central portion that seems to contain the γ, δ and ε chains. The $F_O F_1$ molecules

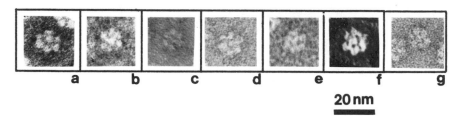

a   b   c   d   e   f   g

20 nm

Fig. 6.16. A gallery of F_1-ATPases from different sources. Top-view projections of electron micrographs of negatively stained preparations of seven different ATPases: (a) from spinach chloroplasts; (b) from beef heart mitochondria; (c)–(g) from various bacteria, including *E. coli* and an archaebacterium. Taken, with kind permission, from G. Schäfer *et al.* (1988) *in* "The Ion Pumps" (W. D. Stein, ed.), pp. 57–66. Alan R. Liss, New York.

can be seen in the electron micrograph to associate side by side to form "strings" [Fig. 6.14(c)] whose dimensions give the width of the F_O portion. The whole F_OF_1-ATPase probably has the structure diagrammed in Fig. 6.15, where the F_O portion, composed of its four chains is transmembranal, while F_1 with its major trigonally arranged α and β units situated in a plane above the membrane, is attached to F_O through the γ, δ, and ε chains. This structure of the F_OF_1-ATPases seems to be a very ancient one. Figure 6.16 shows electron micrographs of F_1-ATPases from a wide range of organisms, including the mitochondria of beef heart, the chloroplasts from spinach, four different eubacteria, and an archaebacterium. The eubacteria split off from the archaebacteria two to three billion years ago. Thus the structure of the F_OF_1-ATPases seems to have been a marvellously persistent one.

6.6.2. Mechanism of Action of the F_OF_1-ATPases

In contrast to the detailed knowledge about the structure of the F_OF_1 system, we are still somewhat poorly off with regard to transport data. There have been few studies of the proton fluxes. The little we know is that the F_O portion can be incorporated into phospholipid bilayers to which it imparts a proton conductivity, inhibitable by DCCD. For the chloroplast enzyme, about 5×10^5 protons are transported per F_O molecule per second, when the transmembrane potential is 100 mV. This is a good deal higher than the rate of a typical carrier or pump (see Table 4.2 in Section 4.2.3). Indeed, translating this flux into a conductance (Section 3.1) gives a value of about 1 pS, a value that is not inconsistent with a channel being the route for the protons (Section 3.4). Significantly, it would seem that subunits of the F_1 portion drastically reduce the proton flux through the F_O portion of the molecule.

We do know a good deal, however, about how the F_0F_1-ATPase acts as an ATPase. A crucial experiment was performed by Paul Boyer, who showed that, even in the absence of a proton gradient, ATP can be synthesized on the enzyme from ADP and phosphate. Thus, the high energy bond in the ATP can be formed, but this ATP does not dissociate readily from the enzyme. Such considerations led Boyer to propose the following reaction scheme for ATP synthesis by mitochondrial F_0F_1 [see Fig. 6.17(a)]. In this reaction scheme E, E^O, and E* are three interconvertible forms of the enzyme that differ in affinities for ADP (symbolized as "D" in the figure) and P_i, and ATP (symbolized as "T"). The form E* binds both ATP (the complex written as E*T) and ADP (and P_i) strongly (written as PE*D), E has a moderate affinity for ADP (and P_i) but little affinity for ATP, while E^O has a moderate affinity for ATP and little for the other two ligands.

Harvey Penefsky obtained experimental data for the reactions depicted in Fig. 6.17(a) at a *single* site on the F_1-ATPase. The reaction was per-

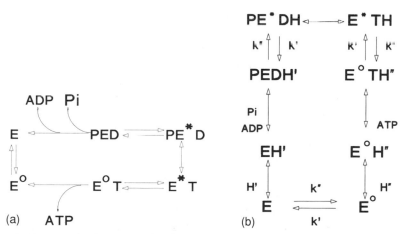

Fig. 6.17. Kinetic schemes for the coupling of ATP synthesis and proton transport in the F_0F_1-ATPases. (a) Reaction scheme for the F_1-ATPase. Binding to E, the free enzyme, by ADP (symbolized as "D") and P_i gives the complex PED. This interconverts to ET ("T" symbolizes ATP), from which ATP can dissociate. The forms labeled with the superscript circle, asterisk, or no symbol are three conformations of the enzyme having, respectively, an affinity for ATP but not ADP, high affinities for both ATP and ADP, and an affinity for ADP but not ATP. (b) Reaction scheme for the integral F_0F_1-ATP synthase that synthesizes ATP and transports protons. As (a) except that, in addition, H' and H" represent protons binding to the pump from sides ' and " of the membrane, respectively. The steps labeled k' and k'' are conformation changes that, in a linked fashion, interconvert the three conformations of the pump enzyme. Figures (a) and (b) taken from W. D. Stein and P. Läuger (1990). *Biophys. J.* **57**, 255–267.

formed at a very low ratio of ATP to enzyme (1:3), i.e., at "subenzyme levels" of the substrate. The enzyme–substrate complex is formed on E^O, and the catalytic step takes place on E^*. The rate-limiting steps are the *release* of the product, from PE^*D and then PED. The crucial observation is that the rate constants for the steps involving the E^* forms are nearly the same in the forward and the backward directions, i.e., ATP bound to the enzyme is nearly in equilibrium with ADP and P_i bound to the enzyme. This is, of course, in great contrast to the situation where these substances are free in solution, when the equilibrium is far over to the side of ATP breakdown. Compensating for the apparent "low-energy" nature of the ATP is the fact that it is *tightly bound* to the E^* form of the enzyme. The dissociation constant for its release from the enzyme is approximately $10^{-12}M$, so that it will seldom be released unless some other circumstance intervenes.

In the fully functioning F_OF_1-ATPase, this circumstance seems to be the transport of proton into the mitochondrion. The chemical reaction depicted in Fig. 6.17(a) can, for the intact F_OF_1 complex, be written (in one possible scheme) as in Fig. 6.17(b), where E is the F_OF_1 complex, and H' and H" represent the protons inside and outside the mitochondrion, respectively. *The flow of protons through F_O is coupled to the release of ATP from its high affinity binding to F_1.* In this way, ATP is delivered to the cytoplasm at the concentration prevailing there, in spite of the fact that ATP is so tightly bound to F_1 when no flow of protons is occurring.

The precise details of the coupling between proton flow and ATP release remain to be established. It would make a fine integration of our picture of coupling in primary active transport if structural evidence could be provided to show that (like the Na^+,K^+-ATPase model of Fig. 6.3) the F_OF_1-ATPase also exists in a number of different conformations, as the model of Fig. 6.17 assumes, with these forms differing in their affinity for ATP, and including an occluded form. One would also have to show that it is the transport of protons that brings about the interconversion between these forms. There is little firm evidence, however, for such a conformation change being at the heart of the action of the F_OF_1-ATPase, and only further research will enable us to reach a full understanding of the system.

We should still discuss four important features of the F_OF_1 system. First, the stoichiometry of transport seems to be three protons being transported for each ATP molecule synthesized. The mitochondrion has a transmembrane potential of about 180 mV, inside negative. This, as we saw in Section 2.2, is equivalent to a concentration gradient of about 10^3. In addition, the pH inside the mitochondrion is about one unit higher than in the cytoplasm, a concentration gradient of 10. The effective gradient across which the protons move is thus about 10^4. With a stoichiometry of 3, the overall transport of the protons involved in the synthesis of a single ATP molecule uses the equivalent of a gradient of 10^{12}, just enough to synthesize an ATP molecule. Second, the mitochondrion uses *this* combination of transmembrane potential difference and the transmembrane concentration gradient. In the chloroplast (which we should note, in passing, has the gradient for protons set up in the opposite direction to that of the mitochondrion) the contribution of the potential is small, that of the concentration gradient large. In bacteria, one or other of these factors can dominate the overall driving force for protons, depending on the circumstances. It would appear that the F_0F_1-ATPase possesses some feature that enables it to readily add up the two components of the overall driving force, the electrical and the chemical gradient of protons. Peter Mitchell has rationalized this in terms of the model of a "proton well" (Fig. 6.18).

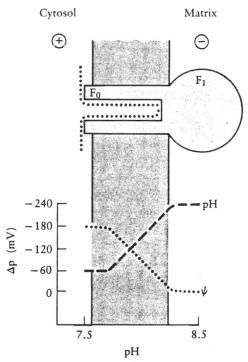

Fig. 6.18. Scheme for a proton well. The upper part of the figure depicts the F_0F_1-ATPase embedded in the membrane. The position of the permeability barrier to protons is indicated by the dotted line. The lower part of the figure shows the chemical gradient of protons through the pump and the electrical potential. Original figure, with kind permission of P. Mitchell, Bodmin, Cornwall.

If the binding sites for the protons transported are deep within the body of F_O, as some evidence indicates, it is possible that a good deal of the transmembrane potential might fall *between the membrane surface and these binding sites*. This would ensure that the protons partition between the bulk medium and these binding sites in accordance with this potential difference. The "well" effectively translates a potential into a concentration. With such a system, it is of no significance whether the electrochemical gradient of protons across the entire membrane is composed mainly of its electrical or its chemical component. In both circumstances, the proton-binding site within F_O will see a far higher concentration of protons that exists in the medium facing the F_1 portion of the pump.

Third, we have not discussed the implications of the fact that there are two binding sites for ATP on each α chain of F_1 and that there are three such units per intact $F_O F_1$-ATPase. The two binding sites on each chain seem to function in a cooperative fashion. We have seen that one ATP molecule binds with extremely high affinity to F_1. The second binds with far lower affinity, about a millionfold less. The result of the binding of this second ATP molecule is a nearly millionfold acceleration in the rate of ATP hydrolysis. Perhaps the function of this second binding site for ATP is regulatory, as we saw for the Na^+,K^+-ATPase. High concentrations of ATP are the signal for the cell that it must reverse the reactions of ATP *synthesis* and the reaction of the hydrolysis is, accordingly, speeded up. The importance of the threefold organization of the F_1 molecule is not at all clear, but since it is preserved universally in living systems, it must be profound. Most intriguing is the suggestion that the ATPase might act in some way as a wheel, rotating on its axis as, first one, then the next, and then the third β subunit, successively go through a cycle of proton transport and ATP synthesis. The bacterial flagellum is a similarly constructed molecule of threefold symmetry that uses the transmembrane proton gradient to drive its rotation in the membrane, suggesting that there may be some real evolutionary link between these two membrane functions!

Fourth, some members of the F-ATPase family (found so far only for certain rather specialized bacterial systems) can use sodium ions rather than protons for the coupling of ion transport to ATP synthesis. This observation suggests very strongly that the proton and sodium ions are playing equivalent roles in the overall process. For both, the synthesis of ATP is brought about by coupling the ion flow to conformational changes in the protein. It seems that it is not any special property of protons (e.g., that they can bind covalently to acid groups on the enzyme) that is needed for the energy coupling.

6.7. THE VACUOLAR PROTON-ACTIVATED ATPase

Both animal and plant cells contain vacuoles whose internal pH is lower than that of the cytoplasm, often by as much as three pH units. The pH is under strict control, the gradient increasing steadily as these vacuoles pass through their developmental histories, from the time that they are newly pinched off from the plasma membrane, through the stages of primary and then secondary endosomes, until they finally become lysosomes. We saw in Section 5.3.6 how such vacuoles are used for the storage of neurotransmitters, these weak bases being trapped within the vacuole by the prevailing low pH. The low pH is maintained by a proton pump, one that is far closer in structure and properties to the F_OF_1-ATPases that we have just discussed than to the vanadate-inhibitable enzymes of Sections 6.1–6.4. There is a family of these vacuolar enzymes or "V-ATPases". Like the F_OF_1-ATPases, they are inhibited by DCCD, and have a complex subunit composition. There is no evidence for the presence of a phosphorylated intermediate during ATP hydrolysis by the V-ATPases.

The stoichiometry of proton transport to ATP split is 2. This should allow such an ATPase to establish an electrochemical gradient equivalent to 6 units of pH (see Box 6.4). Such high gradients are not found in practice, and there is some evidence that the activity of these pumps is regulated by a process of slippage, in which ATP continues to be hydrolyzed although protons are no longer transported through the system.

6.8. AN ATP-DRIVEN ANION PUMP

We saw in Box 6.5 that the development of resistance to drugs in human tumors (and, it has been found, also in parasites) is often due to the selection of mutant strains of the tumor or parasite, as these strains have an increased capacity for pumping out the chemical that is slowing the growth of the nonmutant strains.

Likewise, the bacterium *Escherichia coli* develops resistance to arsenical poisons when it is cultured for a number of generations in the presence of arsenate, arsenite, or antimonate. Barry Rosen and colleagues showed that the development of resistance was due to the expression in these cells of an anion pump, which extrudes oxyanions with the coupled hydrolysis of ATP. The oxyanion pump is composed of two proteins. One is of molecular mass 63,000 Da, is hydrophilic, and has sequences homologous with those of certain nucleotide-binding proteins. The other has molecular mass of 45,500 Da, is hydrophobic, and is presumably the anion-transporting portion. The anion pump is a member of the V-ATPase family.

6.9. BACTERIORHODOPSIN: A LIGHT-DRIVEN PROTON PUMP

In lakes and ponds having high salt content, bacteria can be found (e.g., *Halobacterium halobium*) that have evolved the ability to harness light energy directly into osmotic energy. They do this using a simple yet remarkable proton pump, powered by the absorption of light by a chromophore that is that same pigment, retinal, that the eye uses in the visual process. The retinal–protein complex that performs this light absorption in the eye is known as *rhodopsin*. The analogous protein in the bacterium, that absorbs light and pumps protons, is called *bacteriorhodopsin*. We discussed its structure briefly in Section 1.4 (see Fig. 1.5). Bacteriorhodopsin is a small protein of molecular mass 26,000 Da, containing seven hydrophobic polypeptide sequences that straddle the membrane [see Fig. 1.5(a)]. A purple membraneous patch of bacteriorhodopsin, isolated from the bacterium, is bleached when light shines on it. Walter Stoeckenius, working at the Rockefeller Institute in New York, found that during this bleaching, protons were released from the patch toward the face that was originally at the exterior of the bacterium. If left to itself, the bacteriorhodopsin regenerates its purple color, in the process taking up protons from what was the cytoplasmic surface of the patch. Detailed analysis has shown that the absorption of a photon is followed by a sequence of changes in the spectral properties and hence the structure of the bacteriorhodposin (bR), the sequence closely paralleling that followed by the visual pigment, rhodopsin, when *it* absorbs light! Figure 6.19 shows this sequence and depicts also the change of structure of the pigment retinal when it absorbs a photon, altering in structure from an all-*trans* to a 13-*cis* configuration. The form labeled bR_{570} is the stable form, purple in color and with protons bound to the protein. The transient intermediates K_{590} and L_{550} are also protonated, with the proton now accessible at the outside of the bacterium, while M_{410} is deprotonated. [The subscripts refer to the wavelength (in nanometers) at which the protein maximally absorbs light.] Somewhere between L_{550} and M_{410} the proton is expelled into the medium. It is clear from low-temperature infrared studies that certain carboxyl groups belonging to aspartyl residues of the bacteriorhodopsin are alternately charged and discharged during a cycle of proton pumping.

These data are consistent with the following model of this proton pump (the reader should refer to Figs. 6.19 and 6.20). In the form bR_{570} in Fig.

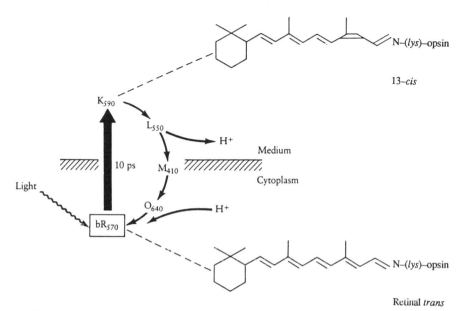

Fig. 6.19. The light-induced cycle of events in bacteriorhodopsin. In the native state of bacteriorhodopsin, the form bR_{570}, the retinal is in all-*trans* conformation as depicted on the lower part of the figure. The molecule absorbs a quantum of light to give the 13-*cis*-retinal (upper sketch in diagram). As the energy so absorbed is released, the conformation of the protein cycles through the forms labeled K_{590}, etc. The subscripts refer to the wavelength (in nanometers) at which the protein maximally absorbs light. Taken, with kind permission, from F. M. Harold, "The Vital Force: A Study in Bioenergetics." Copyright (c) 1986 by W. H. Freeman and Company. Reprinted with permission.

6.19, a proton is bound to the aldehyde–amino link (a so-called Schiff's base) that holds the retinal to a lysine residue of the bacteriorhodopsin (see Fig. 6.20). The charge on this proton suffices to form a salt bridge with the negatively charged carboxyl group of the aspartyl residue labeled Asp-2 in Fig. 6.20. The retinal is in the all-*trans* configuration. When it absorbs a photon, its conformation alters to 13-*cis*, the salt bridge with Asp-2 is broken and the proton carried through the protein (and hence effectively across the membrane) to the region around Asp-1, where a salt bridge is formed once again. This is now the form K of Fig. 6.19. As the protein relaxes through forms L, M, and O the protons are released from Asp-1 to the exterior and further protons are taken up by Asp-2 from the interior. The retinal returns to its all-*trans* configuration, with the establishment of the salt bridge again in conformation bR.

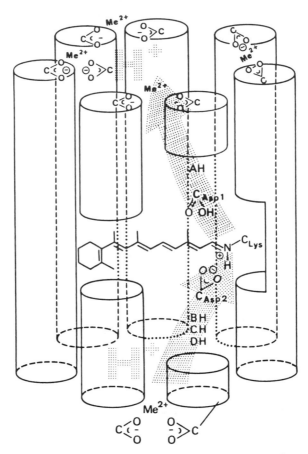

Fig. 6.20.   Path of the proton through bacteriorhodopsin. A tentative scheme for the movement of protons through the protein during the photon-linked isomerization of the protein. The seven cylinders represent the seven transmembrane helices of bacteriorhodopsin. Asp$_1$ and Asp$_2$ are two aspartyl residues that are thought to be successfully protonated and deprotonated during the photocycle and indicate the route of proton transfer through the molecule. The retinal is seen as the horizontal double-bonded structure, forming a Schiff's base with a lysine residue of the protein. Taken, with kind permission, from Gerwert and Hess (1988).

A cycle of light absorption and coupled proton pumping has occurred. What is still, tantalizingly, unclear is the precise location of the aspartyl residues and hence the path that the protons take through the protein. This is the information that we still lack for all carriers and pumps, but with bacteriorhodopsin we seem to be closest to an understanding of the details of membrane transport.

The electrochemical gradient of protons that bacteriorhodopsin establishes with the absorption of light is itself harnessed by the cell in a secondary transport system, an antiport of sodium and protons, whereby sodium is pumped out of the bacterium, maintaining its osmotic balance. The resulting electrochemical gradient of sodium is used in a series of symports (of the type discussed in Chapter 4) to concentrate metabolites within the cell. In addition, this cell possesses a light-driven chloride pump, known as *halorhodopsin*, which pumps chloride into the cell. The interrelationship of these many metabolic systems in the bacterium is only now beginning to be understood.

SUGGESTED READINGS

General

Blostein, R. (1989). Ion pumps. *Curr. Opin. Cell Biol.* **1,** 746–752.
Harold, F. M. (1986). "The Vital Force: A Study of Bioenergetics." Freeman, New York.
Pedersen, P. L., and Carafoli, E. (1987). Ion motive ATPases. *TIBS.* **12,** 186–189.
Stein, W. D. (1986). "Transport and Diffusion Across Cell Membranes", Chapter 6. Academic Press, Orlando, Florida.
Stein, W. D., ed. (1988). "The Ion Pumps: Structure, Function, and Regulation". Alan R. Liss, New York.

Thermodynamics of Pumping

Hill, T. H. (1989). "Free Energy Transduction and Biochemical Cycle Kinetics." Springer-Verlag, New York.
Läuger, P. (1984). Thermodynamic and kinetic properties of electrogenic ion pumps. *Biochim. Biophys. Acta* **779,** 307–341.

Sodium Pump

Apell, H.-J. (1989). Electrogenic properties of the Na,K pump. *J. Membr. Biol.* **110,** 103–114.
Glynn, I. M., and Karlish, S. J. D. (1990). Occluded cations in active transport. *Annu. Rev. Biochem.* **59,** 171–205.
Jorgensen, P. L., and Andersen, J. P. (1988). Structural basis for E_1-E_2 conformational transitions in Na, K-pump and Ca-pump proteins. *J. Membr. Biol.* **103,** 95–120.
Kawakami, K., *et al.* (1985). Primary structure of the alpha-subunit of *Torpedo californica* ($Na^+ + K^+$) ATPase deduced from cDNA sequence. *Nature (London)* **316,** 733–736.
Post, R. L. *et al.* (1969). Flexibility of an active center in sodium-plus-potassium adenosine triphosphatase. *J. Gen. Physiol.* **54,** 306–326.
Shull, G. E. *et al.* (1985). Amino-acid sequence of the catalytic subunit of the ($Na^+ + K^+$) ATPase deduced from a complementary DNA. *Nature (London)* **316,** 691–695.
Skou, J. C. *et al.*, eds. (1988). "The Na^+, K^+-Pump," Parts A and B. Alan R. Liss, New York.

Calcium Pump of Sarcoplasmic Reticulum

Hasselbach, W., and Oetliker, H. (1983). Energetics and electrogenicity of the sarcoplasmic reticulum calcium pump. *Annu. Rev. Physiol.* **45**, 325–339.

MacLennan, D. H. *et al.* (1985). Amino acid sequence of a Ca^{2+}, Mg^{2+}-dependent ATPase from rabbit muscle sarcoplasmic reticulum, deduced from its complementary DNA sequence. *Nature (London)* **316**, 696–700.

Martonosi, A. N. *et al.* (1988). The Ca^{2+} pump of skeletal muscle sarcoplasmic reticulum. *In* "The Ion Pumps: Structure, Function, and Regulation" (W. D. Stein, ed.), pp. 7–14. Alan R. Liss, New York.

Calcium Pump of Plasma Membrane

James, P. H. *et al.* (1989). Structure of the c-AMP-dependent phosphorylation site of the plasma membrane calcium pump. *Biochemistry* **28**, 4253–4258.

Lew, V. L. *et al.* (1982). Physiological $[Ca^{2+}]_i$ level and pump-leak turnover in intact red cells measured using an incorporated Ca chelator. *Nature (London)* **298**, 478–481.

Schatzman, H. J. (1983). The red cell calcium pump. *Annu. Rev. Physiol.* **45**, 303–312.

Gastric H^+, K^+-ATPase

Lorentzon, P. *et al.* (1988). The gastric H^+, K^+-ATPase. *In* "The Ion Pumps: Structure, Function, and Regulation" (W. D. Stein, ed.), pp. 247–254. Alan R. Liss, New York.

Polvani, C., Sachs, G., and Blostein, R. (1989). Sodium ions as substitutes for protons in the gastric H,K-ATPase. *J. Biol. Chem.* **264**, 17854–17859.

Sachs, G. *et al.* (1982). The ATP-dependent component of gastric acid secretion. *In* "Electrogenic Ion Pumps" (C. F. Slayman, ed.), Curr. Top. Membr. Transp., Vol. 16, pp. 135–159. Academic Press, New York.

Multidrug Resistance

Gottesman, M. M., and Pastan, I. (1988). The multidrug transporter, a double-edged sword. *J. Biol. Chem.* **263**, 12163–12166.

F_0F_1 ATPases

Boyer, P. D., and Kohlbrenner, W. E. (1981). The present status of the binding-change mechanism and its relation to ATP formation by chloroplasts. *In* "Energy Coupling in Photosynthesis" (B. R. Selman and S. Selman-Reimer, eds.), pp. 231–240. Elsevier/North-Holland, Amsterdam.

Gräber, P. (1987). Primary charge separation and energy transduction in photosynthesis. *In* "Bioelectrochemistry II" (G. Milazzo and M. Blank, eds.), pp. 379–429. Plenum, New York.

Penefsky, H. S. (1985). Energy-dependent dissociation of ATP from high affinity catalytic sites of beef heart mitochondrial adenosine triphosphatase. *J. Biol. Chem.* **260**, 13735–13741.

Senior, A. E. (1988). ATP synthesis by oxidative phosphorylation. *Physiol. Rev.* **68**, 177–231.

Vacuolar and Anion Pumps

Moriyama, Y., and Nelson, N. (1988). The vacuolar H^+-ATPase, a proton pump controlled by a slip. *In* "The Ion Pumps: Structure, Function, and Regulation" (W. D. Stein, ed.), pp. 387–394. Alan R. Liss, New York.

Rosen, B. P. *et al.* (1988). Molecular characterization of a unique anion pump: the ArsA protein is an arsenite (antimonate)-stimulated ATPase. *In* "The Ion Pumps: Structure, Function, and Regulation" (W. D. Stein, ed.), pp. 105–112. Alan R. Liss, New York.

Bacteriorhodopsin

Butt, H. J. *et al.* (1989). Aspartic acids 96 and 85 play a central role in the function of bacteriorhodopsin as a proton pump. *EMBO J.* **8**, 1657–1663.

Engelhard, M., *et al.* (1985). Light-driven protonation changes of internal aspartic acids of bacteriorhodopsin: An investigation by static and time-resolved infrared difference spectroscopy using [$4^{13}C$] aspartic acid labeled purple membrane. *Biochemistry* **24**, 400–407.

Gerwert, K., and Hess, B. (1988). Light induced intramolecular processes of bacteriorhodopsin monitored by time resolved FTIR (Fourier transform infrared) spectroscopy. *In* "The Ion Pumps: Structure, Function, and Regulation" (W. D. Stein, ed.), pp. 321–326. Alan R. Liss, New York.

Stoeckenius, W., and Bogomolni, R. A. (1982). Bacteriorhodopsin and related pigments of Halobacteria. *Annu. Rev. Biochem.* **51**, 587–616.

CHAPTER 7

Regulation and Integration of Transport Systems

The many transport systems that we discussed in Chapters 1–6 do not exist independently in the cell. Instead, many of them are subject to *regulation* by factors within the cell that stimulate or inhibit them in response to specific signals that indicate some required change in cell properties. Some of the many ways in which regulation is achieved in the cell are as follows:

1. The most obvious way, and a very frequent one, is by the control of the *amount* of transporter present in the membrane.

(a) New transporter molecules can be synthesized by the cell in response to a specific signal, and then inserted into the membrane. The lactose symporter, discussed in Section 5.3.6, is the classic case of a transporter whose synthesis is switched on when the bacterium faces the need to accumulate its substrate. Regulation of the sodium pump by increased synthesis of the protein is discussed in Section 7.2.2.

(b) In other cases, the transporters exist in preexisting form *within the cytoplasm* in specialized storage systems, from which they are released by appropriate signals and inserted into the cell membrane. We saw an example of this in Box 4.1, where we considered the regulation of the rate of glucose transport across cell membranes by the hormone insulin. Later in this chapter, we shall see that water channels are moved from a cytoplasmic store to the cell membrane, when certain body cells augment their ability to transport water, in response to specific, hormonal signals. When transport is to be reduced again, the transporters are selectively removed from the membrane and can be returned to the store.

2. Transporters existing in the membrane can be *activated or inhibited* by being altered chemically, in response to enzymatic changes within the cell, these changes themselves being brought about as a response to a specific signal. Transporters can be phosphorylated or dephosphorylated by kinases and phosphatases acting from the cytoplasmic phase. The $Na^+ + K^+ + 2Cl^-$ cotransporter, discussed in Section 5.3.6 and later in this chapter, is an example of a system regulated indirectly by a kinase.

3. In many cases, the activity of a transporter is regulated by *the level of its substrate*. We shall see later in this chapter how a sodium channel in the kidney is switched off when sodium binds to a regulatory site on the channel. We saw in Section 6.1.1 that the affinity of the sodium pump for potassium is determined by the concentration of ATP.

Many of the transport systems operate in an *integrated* fashion, with two or more such systems acting in concert to set the level of a solute within the cell or to transport the solute across the cell, in at one face and out at another, or to sequester a solute within an intracellular organelle. Thus, in some of the cases discussed in Section 5.3, the cotransport of sugars and amino acids carries these substrates into the cell at one face, while they leave it at the other face by a facilitated diffusion system. In Section 5.3.6, we saw how a neurotransmitter was accumulated within a storage granule within platelet cells by a combination of two systems, acting in tandem.

In this chapter we shall discuss a few examples of regulation (focusing particularly on the transport systems concerned in the regulation of the volume of the cell) and a few cases of integration (where it is the kidney as an organ that is the object of our attention). We have chosen these particular cases as examples of the many systems that are being actively researched at this time.

7.1. REGULATION OF CELL VOLUME

We saw in Section 3.8 that the presence of impermeant anions within a cell makes it perpetually vulnerable to osmotic flooding. Animal cells parry this threat by the continual pumping out of incoming sodium (the double Donnan effect). Long-term regulation of cell volume is achieved in this fashion. It is obviously an advantage to the cell to invest the minimum of metabolic energy in this volume regulation. Hence, there will be an evolutionary trend to select for a plasma membrane with as low a permeability as possible for the ion that is pumped out across it. Thus the long-term regulatory systems operate in the face of a very low permeability to sodium and tend, therefore, to be slow and ponderous. Cells also require mechanisms for the *acute* regulation of cell volume in response to rapid changes in external osmolarity. This acute regulation of volume decrease and of volume increase is performed by a variety of mechanisms that include the controlled switching on and off of the co- and countertransport systems for Na^+, K^+, and Cl^-, and channels for K^+ and Cl^-.

If cells find themselves in a *hypersomotic* environment, in which they would shrink, they maintain volume by taking up the major ions in the environment, sodium and chloride, these being accompanied by water, which, entering the cell osmotically, restores the original volume. The process is known as *regulatory volume increase* (RVI). Different cells, as we shall see, use different means to achieve the uptake of sodium ions. If cells are subject to sudden flooding on being placed in a *hypoosmotic environment*, they lose the major internal ion, potassium, again by a number of different mechanisms. The loss of potassium is accompanied by an osmotic flow of water, and the cell volume is restored. This process is termed *regulatory volume decrease* (RVD). The rapid increases in ion permeability that bring about the regulatory volume changes are transient, and eventually long-term regulation comes into play for the maintenance of volume stability. We shall describe first the factors that fix the volume of the cell in the long term and then discuss the short-term controls that regulate cellular volume.

7.1.1. Long-Term Determination of Cell Volume

In both the short term and the long term, the cell volume is determined by a balance between the pumping out of ions and their leaking back into the cell. A beautifully simple equation [Eq. (B7.1.5)], derived in Box 7.1, describes this process. (Jakobsson's (1980) paper gives a full treatment.)

Box 7.1. The Post–Jolly Equation Relating Cell Volume, Cell Content, and the Pump–Leak Ratio

An equation that links the transport properties of cells with their equilibrium volume was formulated by the biochemists, Post and Jolly. We shall develop their equation in stages, taking first the very simple case where there are only two solutes present, neither of them charged.

Consider Fig. 7.1(a), which depicts a cell suspended in a medium containing only a permeant molecule at concentration $[P]_e$. Within the cell is an impermeant molecule at concentration $[A]_i$ and the permeant

---

(Continued)

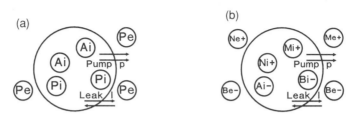

Fig. 7.1. Pump–leak relations in a schematic cell. (a) The scheme for nonelectrolytes: P is a permeable species, A an impermeable one. The subscripts i and e refer to the intracellular and extracellular species; p and l are the rate constants for the pump and the leak processes. (b) The scheme for electrolytes: M, N, B are permeable, A impermeable; A and B are anions, M and N cations.

molecule at concentration $[P]_i$. At osmotic equilibrium, we have

$$[A]_i + [P]_i = [P]_e \qquad (B7.1.1)$$

Let us assume that there is a pump that pumps P out of the cell and operates with a rate constant (i.e., pumping rate divided by concentration) k_p. P leaks in and out of the cell across the membrane according to leak permeability constant, k_l. At steady-state, the total efflux of P (pump + leak), must equal its total influx (leak only), so that we have

$$k_l [P]_i + k_p [P]_i = k_l [P]_e \qquad (B7.1.2)$$

Solving these two equations for $[A]_i$, we obtain

$$[A]_i = \frac{[P]_e}{(1 + k_l/k_p)} \qquad (B7.1.3)$$

This shows us how the concentration of $[A]_i$ will depend on $[P]_e$ and the *leak-to-pump ratio* k_l/k_p. Now, $[A]_i$ is simply the *amount* of internal impermeant material divided by the equilibrium cell volume, V. Write X_i for the amount of this internal impermeant material. Then we have

$$[A]_i = X_i/V \qquad (B7.1.4)$$

Substituting in Eq. (B7.1.3), and rearranging, gives us the Post–Jolly equation:

$$V = (X_i/[P]_e)(1 + k_l/k_p) \qquad (B7.1.5)$$

(Continued)

Equation (B7.1.5) states that the volume at equilibrium of an osmotically responsive cell will be determined by the amount of impermeant material (X_i), the concentration of extracellular material ($[P]_e$) and the leak-to-pump ratio (k_l/k_p), and by these three parameters alone. Of these values, $[P]_e$ is clearly a given datum of the natural world and is outside the control of the cell in question. The quantity X_i must be fixed by the biological requirements of the cell, i.e., by the amount of impermeant matter (enzymes, nucleic acids, metabolites, etc.) it must contain in order to carry out its various functions. The leak/pump ratio must, therefore, be selected for, in order to arrive at a cellular volume suitable for the cell in question.

Take now a more realistic model for the cell, that depicted in Fig. 7.1(b). Here the cell contains two cations (M^+ and N^+), only one of which is pumped out of the cell, an impermeant anion (A^-) and a permeant anion (B^-). The concentrations of the various species are shown on the figure. We have now to take into account the osmotic equilibrium, the Donnan relation for the permeant species due to the presence of the impermeant species A^-, and the eventual attainment of the steady-state, as M^+ is pumped out by the pump and leaks back through the leak pathway. Solution of the relevant equations shows that the volume is given by

$$V = (X_i/[M^+]_e)(1 + k_l/k_p) \qquad \text{(B7.1.6)}$$

Equation (B7.1.6) is exactly equivalent to Eq. B7.1.5, with the extracellular concentration of the pumped ion replacing the external concentration of the pumped molecule. The Donnan condition applies, so that

$$[B^-]_i/[B^-]_e = [N^+]_e/[N^+]_i = r \qquad \text{(B7.1.7)}$$

where r is the Donnan ratio of Eq. (3.5) of Section 3.7, and is always less than unity.

Finally, for the pumped ion M^+, we have

$$[M^+]_i = [M^+]_e + [N^+]_e(1 - 1/r) \qquad \text{(B7.1.8)}$$

Thus $[M^+]_i$, the intracellular concentration of the ion that is pumped out, is, of course, *less* than $[M^+]_e$, its concentration in the extracellular medium by an amount determined by the Donnan ratio, itself determined by the amount of intracellular impermeant material.

The equations derived in Box 7.1 give a simple description of the forces that control the volume of the cell. The amount of impermeant intracellular matter, $[X^-]_i$, the leak/pump ratio of the pumped-out ion (k_l/k_p), and concentration of the latter in the extracellular fluid, $[M^+]_e$, (as this is *not* under the control of the cell), determine the cell volume. The cell volume, in turn, determines the concentration of impermeant anion, and this fixes the Donnan ratio and hence the equilibrium concentrations of all non-pumped ions. The results are remarkably general, yet simple! The question then arises as to what determines the leak/pump ratio in the resting cell and in the cell subjected to acute changes in cell volume? Any mechanism for the regulation of cell volume requires that the cell can measure its own volume. We do not know how this is done, nor do we know how the cell has achieved the balance of volume and leak/pump ratios that are involved in long-term volume regulation. We are, however, beginning to know something about how leaks and pumps are controlled in the short-term responses.

7.1.2. Short-Term Regulation of Cell Volume

When a cell is taken from the isotonic medium in which it is living and placed in a hypotonic (i.e., less concentrated) solution, it immediately swells as water enters the cell down its osmotic gradient. (Note the rapid *upward* deflection in Fig. 7.2, first half of record.) This flow of water is extremely rapid, being completed within a minute in even the largest cells. In many types of cells, the lymphocytes depicted in Fig. 7.2 are a well-studied example, and the initial swelling is followed by a far slower shrinkage, the slow *downward* deflection of the curve in Fig. 7.2, until the original volume is restored. This is acute *regulation* of volume. The cells at their original volume can now be returned to the original isotonic medium, with which they are no longer in osmotic equilibrium (right half of record in Fig. 7.2). The cells shrink, again extremely rapidly, and then swell again far more slowly as the regulatory systems are activated, restoring the original volume over a 10-min period.

For all these systems, cell volume must, however, in the final analysis, be determined by the Post–Jolly equation that was derived in Box 7.1. When volume regulation occurs, the volume returns to its original value in the face of a changed value of the extracellular tonicity $\{[M^+]_e$ in Eq. (B7.1.6)$\}$. Since the total amount of impermeant intracellular matter is probably not altered (although cases are known, in insects, for example, where this is indeed the basis of regulation), it must be the ratio of leak/pump rate constants (k_l/k_p) that is being affected in acute volume regulation. It turns out that much of the volume regulation is due to changes in

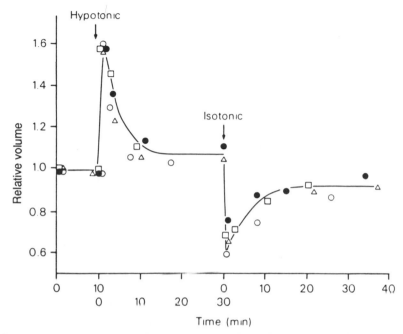

Fig. 7.2.   Volume regulation in lymphocytes. Volume changes were measured in periph eral blood lymphocytes in isotonic saline at 37°C. At the first arrow 300 mos$M$ (milliosmolar) of sucrose or saline was added to the suspension medium. At the second arrow, the cells were resuspended in the original isotonic medium. Reproduction from S. Grinstein *et al.* (1983). *J. Gen. Physiol.* **82,** 619–638.

the leak rate constant. Permeabilities for various ions alter remarkably, as we shall see, and in accordance with Eq. (B7.1.6), these bring about the acute volume regulation.

7.1.2.1. REGULATORY VOLUME DECREASE

Regulatory volume decrease (RVD) occurs when, as in the first part of the curve in Fig. 7.2, cells are placed in a hypotonic (diluted) medium, swell initially, and then gradually regain their initial volume. To do this, they have to extrude the major intracellular ion, generally potassium, accompanied by chloride. The driving force for the extrusion of potassium is simply the electrochemical gradient of that ion. Recall from Section 3.7 [Eq. (3.5)] that the Donnan ratio helps fix the ratio of intracellular to extracellular potassium and, if the concentration of impermeant material within the cell is high, determines this at a large number. If the extracellular medium is diluted, potassium will flow out of the cell to restore the ratio given by the Donnan condition. The mechanism by which it flows out differs in different cell types.

In the lymphocyte preparation depicted in Fig. 7.2, it would appear that potassium and chloride both leave the cell through swelling-activated channels, each ion through its own specific channel. The evidence for this view is that known blockers of potassium channels can inhibit the cell shrinkage and inhibit the potassium efflux activated by swelling. Similarly, inhibitors of chloride channels inhibit cell shrinkage and inhibit the chloride flux. One can block potassium channels with the inhibitor quinine, thus blocking cell shrinkage, and then overcome the block to shrinkage by adding gramicidin, the bacterially derived potassium channel (see Section 3.1 for more on gramicidin). Chloride ions leave the cell by their own channels, as patch-clamping confirms. The mechanism of activation of the potassium channels seems to involve release of intracellular calcium. Swelling-induced activation is prevented if intracellular stores of calcium are depleted before swelling takes place.

In contrast, in duck red blood cells and in the giant red blood cells of the amphibian *Amphiuma means*, carrier systems, rather than channels, seem to be involved in the potassium loss that follows cell swelling. In duck red cells, a potassium–chloride cotransport system has been postulated, while in *Amphiuma* cells, the evidence is that a potassium–proton antiporter is the system that is activated by cell swelling. How an antiporter can be harnessed to bring about a *net* transport of osmotically active material is discussed in the next section (see Fig. 7.4).

Whatever the mechanism of the immediate potassium loss, our analysis of the Post–Jolly equation (Box 7.1) must lead us to the conclusion that the *long-term* maintenance of the cell volume at its original level, when the external medium is reduced in concentration, must involve one of the parameters of Eq. (B7.1.6). The potassium or chloride permeabilities, which are the ions that are not *pumped* out of the cell, do not come into this equation. It is most likely that it is k_p, the rate constant of sodium pumping, that is increased in the long term. This increase results from the lowering of the intracellular potassium concentration which reduces the competition with sodium for efflux on the sodium pump.

7.1.2.2. REGULATORY VOLUME INCREASE

Regulatory volume increase (RVI) can be seen in the right-hand half of the record in Fig. 7.2. In all cases of RVI, there is an increased flow of sodium into the cells, the primary driving force for which is the electrochemical gradient of sodium, maintained by the sodium pump, but the mechanisms by which RVI occurs are diverse.

In the lymphocytes of Fig. 7.2 it seems clear that the primary event is an *exchange* of Na^+ and H^+ by the sodium–proton antiporter (Section 5.2.5) of those cells. Figure 7.3(a) shows that sodium has to be present in the extracellular medium if cell swelling is to occur, Fig. 7.3(b) shows that

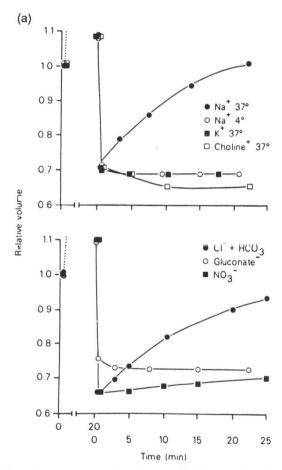

Fig. 7.3. Effects of temperature and cation composition on volume regulation in lymphocytes. (a) The cells were first equilibrated in half-isotonic saline (dotted line). They were then brought into isotonic media and the effect of (top figure) temperature and cations and (lower figure) anions on the volume regulatory increase, measured as indicated. (b) The uptake of sodium was measured in cells that had been preequilibrated in half-isotonic saline (solid symbols) or left undisturbed (open symbols), with (squares) or without (circles) amiloride, an inhibitor of the Na/H exchanger. (c) The change in pH, measured fluorometrically, when cells are made to shrink. The controls show the absence of alkalinization when sodium is absent extracellularly or amiloride is present. All three figures taken, with kind permission, from S. Grinstein *et al.* (1985). *Mol. Physiol.* **8**, 185–189.

(b)

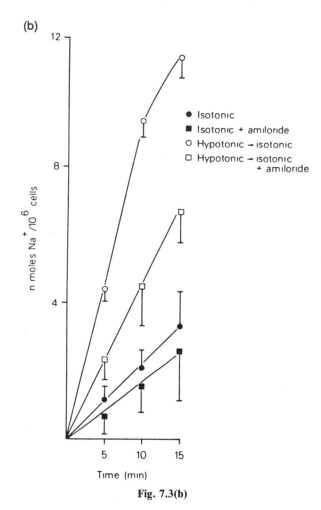

Fig. 7.3(b)

refer to a cell biology textbook). The increase in cellular cyclic AMP leads to phosphorylation of sites on a protein associated with the cotransporter and this to a switching on of its activity.

   Finally, in Ehrlich ascites tumor cells, entry of sodium following cell shrinkage seems to be brought about by a sodium–chloride cotransporter. The system is inhibited by bumetanide, but it is clear that it does not carry potassium into the cells and that its stoichiometry is $Na^+:Cl^- = 1:1$.

sodium enters the cell, while Fig. 7.3(c) shows that the intracellular pH rises as lymphocytes swell, and that acid appears in the extracellular medium. The compound amiloride, a well-known inhibitor of the Na^+/H^+ antiporter (see Section 5.2.5), blocks cell swelling [Fig. 7.3(b)], strongly suggesting that this antiporter is involved. It is clear, however, that an ion-for-ion *exchange* of sodium for proton would not in itself cause a change in the osmotic content of the cell. The loss of protons must be nullified insofar as its contribution to the internal osmotic material is concerned. Part of the protons are restored by release from intracellular buffering groups, but this can contribute only a small part of the outgoing proton. Part is released from bicarbonic acid present originally within the cell as dissolved carbon dioxide, but the major part is removed together with bicarbonate ions that leave the cell in exchange for chloride ions. Thus the parallel antiporters of Na^+/H^+ and Cl^-/HCO_3^- act in concert (Fig. 7.4) to bring about the net import of sodium chloride, with its osmotic contribution, against a net production of bicarbonic acid intracellularly from water and gaseous carbon dioxide, neither making any osmotic contribution. The volume increase is not only blocked by blockers of the Na^+/H^+ antiporter such as amiloride but is also reduced markedly by an inhibitor of Cl^-/HCO_3^- exchange, the compound DIDS (see Section 4.8.1).

How is the activation of the Na^+/H^+ antiporter itself brought about? Recall our study of this system in Section 5.2.5. It is the system activated by many growth factors, putting the cell in a new state in which it gets ready for cell division. Remember, too, that the system at rest is inhibited at the ambient intracellular pH and that activation shifts its pH–activity curve to higher pH values. (This was discussed in Section 5.2.5.) Thus, after activation by growth factors it is able to operate at the same ambient pH. In the same way, cell shrinkage acts by shifting the pH–activity curve of the antiporter to more alkaline regions, so that the system is switched on although the intracellular pH does not alter.

In contrast, in the well-studied duck red cell system, shrinkage-activated entry of sodium occurs on the $Na^+ + K^+ + 2Cl^-$ cotransporter that we studied in Section 5.3.6. The swelling response is blocked if the extracellular medium lacks sodium, or potassium, or chloride ions. The entry of sodium and potassium and chloride ions can all be measured as a response to cell shrinkage. Cell swelling is blocked in these cells by the addition of bumetanide or furosemide, known specific inhibitors of the $Na^+ + K^+ + 2Cl^-$ cotransporter.

Activation of the cotransporter can be achieved by treatment of the cells with catecholamines, drugs that bind to specific receptors on avian cell membranes and activate the cyclic AMP system of the cell (for cAMP

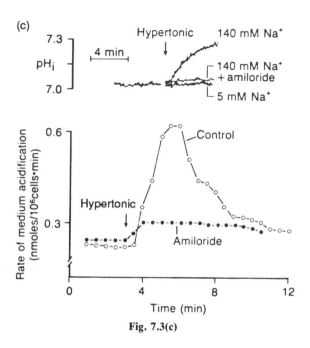

Fig. 7.3(c)

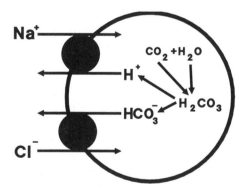

Fig. 7.4.   Two parallel transporters that, in concert, act so as to bring NaCl into the cell. The heavy black circles represent the Na^+/H^+ (top) and the Cl^-/HCO_3^- (bottom) antiporters. The synthesis (or dehydration) of H_2CO_3 catalyzed by carbonate dehydratase occurs within the cell (large circle).

7.2. INTEGRATION OF TRANSPORT SYSTEMS

7.2.1. Epithelia, with Special Reference to the Kidney

We turn now to a new topic, the integration of transport systems. We will consider in detail a range of cell types in which opposite faces of the cell possess membranes that contain *different transport systems*. All of the cell types that we will discuss are present in *epithelia*, and we shall start with a brief description of the morphology of these tissues.

7.2.1.1. MORPHOLOGY OF EPITHELIA

An epithelium is a sheet or pavement of cells in which each cell is joined to its neighbor by a *tight junction*. Such a sheet of cells may be flat, as in a skin or bladder, or be arranged in a tubular cylinder, as in the tubules of a kidney or an intestine. In kidney tubules, the cells generally have the shape of a truncated cone [Fig. 7.5(a)], with the wide base of the cone and its sides forming what is referred to as the *basolateral surface* of the cell, while the apex of the cone is known as the *apical surface*. The basolateral surface always faces the blood supply of the tissue (and is thus sometimes referred to as the *serosal surface* (compare "serum"). In some tissues (amphibian skin, intestine) the apical surface is covered with a mucus secretion, and is termed the *mucosal surface*. Where the apical surface faces into the lumen (the space within a tubule), it is often referred to as the *luminal surface*. To stress the generality of our description, in what follows we shall only use one set of terms, the apical and basolateral surfaces. The tight junctions hold the cells in the sheet close to one another, forming a collar that cuts off the apical from the basolateral surfaces [Fig. 7.5(b)].

A range of techniques has been used to investigate the transport properties of epithelia. Easiest to study are flat sheets such as skin, or those tissues (such as intestine, gallbladder or urinary bladder) that can be cut open to form flat sheets. Such a sheet can be mounted in the Ussing chamber discussed in Section 5.3.1 [Fig. 5.7(a)], to form a barrier between two baths whose composition can be controlled.

Electrical measurements can be made of the current flowing across the sheet, and the media bathing the two faces of the sheet can be held at any required potential difference. Finally, chemical measurements can be made of the amount of any substance that flows across the epithelium. To study tubules is more difficult. The sophisticated techniques used often require the dissection of the tubules from an organ such as the kidney, the insertion of electrodes into the lumen of the tubule, and the isolation of a portion of the tubule using oil droplets (Fig. 7.6). With these methods

(a)

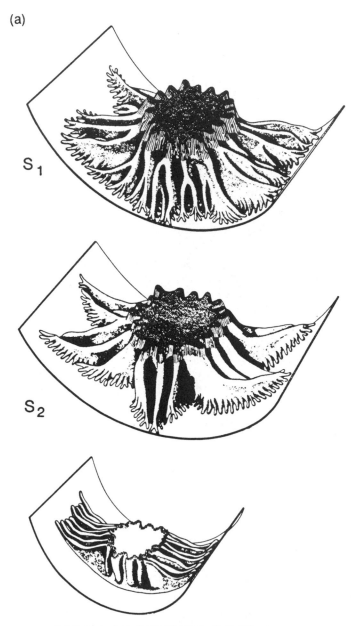

S_1

S_2

THICK ASCENDING LIMB

Fig. 7.5 (*Figure continues*)

(b)

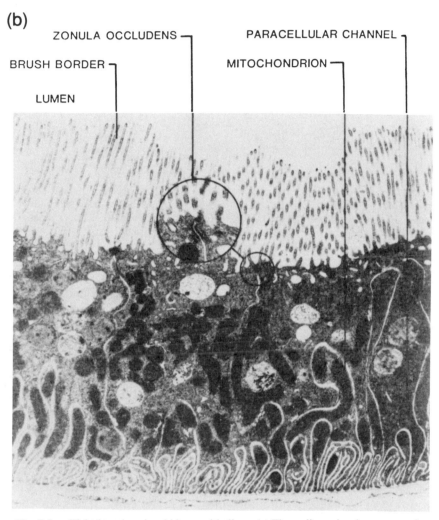

Fig. 7.5. Tight junctions in a kidney epithelium. (a) Three-dimensional reconstructions of cells from S_1 and S_2 segments of proximal tubule and thick ascending limb (See Fig. 7.8). The brush border is uppermost. Below this lies the basolateral membrane, which folds and digitates to form fingers that overlie the basement membrane. The tight junction region [see (b)] is not shown. (b) A transmission electron micrograph of a portion of the proximal tubule of rabbit kidney. The brush border is uppermost, as in (a). The circular inset is a higher magnification of the tight junction region, termed here the *zonula occludens*. (Taken, with kind permission, from L. Welling, VA Medical Center, Kansas City.)

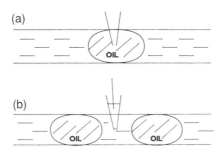

Split Oil-drop

Fig. 7.6.   The split oil-drop technique. In (a), an oil drop has been inserted into a kidney tubule and a micropipette inserted into the drop. In (b) the drop has been split into two, sealing off a portion of the tubule, the contents of which can be sampled and across the membrane of which, potentials and current flows can be measured.

available, one can perform all the transport experiments that have been discussed in previous chapters, but now measuring fluxes and electrical currents across the epithelium, rather than across the membrane of an individual cell. These fluxes can be measured as a function of the chemical composition of the two different media bathing the two faces of the sheet, the apical and the basolateral surfaces.

7.2.1.2. TIGHT, INTERMEDIATE, AND LEAKY EPITHELIA

The various types of epithelia can be classified into groups, depending on the combination of water permeability and electrical resistance that they display. Table 7.1 lists some epithelia classed into three levels of permeability resistance, the three groups being termed tight, intermediate, and leaky epithelia, respectively. As might be expected, the three groups of epithelia have different *functions* in the body. It is somewhat, but not too much, of an oversimplification to say that tight epithelia function as an ion- and water-tight barrier within the body (as in the case of the collecting duct of the mammalian kidney) or between the body and its surroundings (as in amphibian skin). As we shall see, the barrier properties of a tight epithelium can be *regulated*, when necessary, so as to bring about a directed flow of ions and water. Intermediate epithelia allow a greater flow of ions but are still not highly permeable to water. Thus they can separate ions and water. Finally, leaky epithelia are highly permeable to many ions and to water, but not to the larger solutes. They function as selective valves, filtering out substances required by the body from those that must be expelled from it. It is largely the tight junctions between the

TABLE 7.1
Some Epithelial Tissues Classified

Epithelium type	Trans-epithelial resistance (Ohms/cm^2)	Water permeability	Postulated function
"Tight" epithelia			
Frog skin[a]	1,300	Low, increased by anti-diuretic hormone	Defence against dehydration, salt loss
Toad urinary bladder[b]	1,000–3,500	Low, increased by anti-diuretic hormone (see Table 2.6)	Salt retention, water storage
Rabbit urinary bladder[c]	5,000–10,000		
Turtle colon[d]	700		Salt absorption
Rabbit, kidney collecting duct[e]	270	Low, increased by anti-diuretic hormone	Water retention in dehydration
"Intermediate" epithelia			
Thick ascending limb of kidney, mammals[f]	27	Low	Dilution of urine
Early distal tubule amphibian kidney	60	Low	Dilution of urine
Coprodeum of the hen cloaca[g]	150–250	Low	Dilution of urinary fluid
Submaxillary gland main duct[h]	7.4	Low	Dilution of salivary fluid
"Leaky" epithelia			
Gall bladders[i]			
Roach	110	High	Water removal, lead-
Frog	130	High	ing to concentra-
Goose	30	High	tion of gall fluid
Rabbit	23	High	
Ileum, Rabbit[j]	55	High	Solute absorption
Jejunum, Rat[k]	30	High	Solute absorption
Proximal tubule, kidney, Rat[k]	6	Very high (see Table 2.6)	Massive solute reabsorption

[a] W. Nagel (1980). *J. Physiol. (London)*, **302**, 281–295.

[b] MacKnight *et al.* (1980).

[c] Lewis, S. A., Eaton, D. C., and Diamond, J. M. (1976). *J. Membr. Biol.* **28**, 41–70.

[d] Thompson, S. M., and Dawson, D. C. (1978). *J. Gen. Physiol.* **72**, 269–282.

[e] Jamison and Kriz (1982).

[f] Greger, R., and Schlatter, E. (1983). *Pflügers Arch.* **396**, 325–344.

[g] Bindslev, N. (1981). *in* "Water Transport across Epithelia", H. H. Ussing *et al.* (eds) Munksgaard, Copenhagen, pp 468–481.

[h] Van Os, C. H. *et al.* (1981). *in* "Water Transport across Epithelia," H. H. Ussing *et al.* (eds) Munksgaard, Copenhagen, pp 178–187.

[i] Reuss, L. (1978). *in* "Membrane Transport in Biology, Vol IV B", G. Giebisch *et al.* (eds), Springer-Verlag, Berlin, pp 853–898.

[j] Holman, G. D. *et al.* (1979). *J. Physiol. (London)* **290**, 367–386.

[k] Fromter and Diamond (1972).

cells of the various epithelia that give the sheet as a whole a high or intermediate degree of leakiness. The ions and water that flow across a leaky epithelia in part flow *between the cells* of the sheet rather than through them. This route is referred to as the *paracellular* route for ion or water transport.

Box 7.2. In Cystic Fibrosis, the Regulation of an Epithelial Chloride Channel Is Defective

Cystic fibrosis (CF) is the most common lethal genetic disease among Caucasians. It affects about 1 in 5000 in such populations and gives rise to major malfunctioning of mucous epithelia such as the sweat glands and the linings of the lungs and airways. Quinton showed that the defect was associated with an abnormally low transepithelial chloride transport in the cystic fibrosis patients. Patch-clamp studies (see Section 3.4.1) have definitely localized the defect in CF to the failure of the chloride channels at the apical surface of these epithelia. Figure 7.7(a) depicts the pathway for chloride movement in the sweat

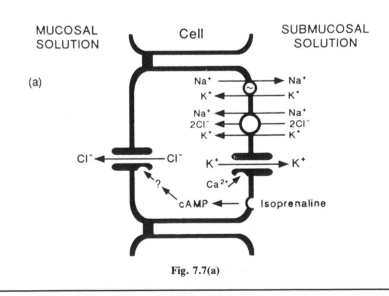

Fig. 7.7(a)

(Continued)

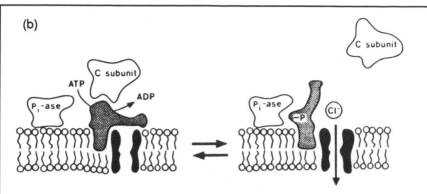

Fig. 7.7. Pathway for chloride in the airway epithelium. (a) A model of the pathway for chloride across the airway epithelium. The $Na^+ + K^+ + 2Cl^-$ cotransporter is depicted as a large circle, the sodium pump as a smaller circle, and the chloride and potassium channels as thick, paired parallel lines. Tight junctions close off the paracellular route between adjacent cells. A scheme for the activation of the chloride channel by epinephrine is shown in the lower part of the figure. Reprinted by permission from M. J. Welsh and C. M. Liedtke, *Nature* **322,** 467–470. Copyright © 1986 Macmillan Magazines Ltd. (b) A scheme for the regulation of the chloride channel. Phosphorylation of one of its protein subunits (the hatched portion) opens the channel. Its dephosphorylation is brought about by an adjacent phosphatase. Reprinted by permission from R. A. Schoumacher *et al.*, *Nature* **330,** 752–754. Copyright © 1987 Macmillan Magazines Ltd.

gland and airway epithelia. Chloride ions enter the cells via the $Na^+ + K^+ + 2Cl^-$ cotransporter situated in the basolateral surface and leave the cell through the apical chloride channels. These are relatively specific for chloride, preferring it over sodium to the extent of about 7.5 to 1. When the chloride channels are studied within patches that have been excised from the cell membrane, there is no difference in the behavior of normal and CF tissues. Both contain effective chloride channels. When patches are studied with the electrode in situ, the "cell-attached" mode, a marked difference between normal and CF cells becomes clear. Normal cells increase their chloride channel activity when they are stimulated by hormones such as adrenaline, while CF cells do not. Yet the CF cells contain chloride channels in normal amounts and, as we saw, these function otherwise normally when excised.

Recent research in the laboratories of Frizzell and of Welsh has shown that the chloride channels of normal cells are capable of being

(Continued)

stimulated even when excised, if they are treated with an enzyme system that phosphorylates cell membranes. Excised patches from such cells, when cyclic AMP is added together with ATP and the catalytic subunit of the cAMP-dependent protein kinase (see Section 3.6.4), show a pronounced increase in chloride channel activity; the lower part of Fig. 7.7(a) depicts this stimulation diagrammatically. Patches made from the membranes of CF cells fail to show such stimulation. Thus they are defective in the *regulation* of their chloride channel activity. Just how this defect arises is the center of much research activity at present. It could be that the channel itself is unable to undergo phosphorylation. An alternative possibility is that it is indeed phosphorylated but fails to undergo the conformation change [depicted in Fig. 7.7(b)] that allows it to act as a passage for chloride. The recent cloning and sequencing of the gene that is defective in CF should lead to substantial progress in this area.

We can understand how these different epithelial types act, and act in coordinated fashion in the body, by considering at a simple level the mechanism of action of the mammalian kidney, the transport organ par excellence.

7.2.1.3. THE MAMMALIAN KIDNEY

While this chapter is not at all meant to be a description of how the kidney functions (for more information on this subject the texts listed in the bibliography should be consulted), we need some background in order to appreciate the exquisite functional adaptation of the kidney. Figure 7.8 depicts a simplified view of a *nephron*, the unit of function of a kidney. (There are about one million of these in each of our kidneys!) An ultrafiltrate of the blood, containing all the ions and small solutes but none of the proteins of that fluid, is forced out of a capillary tuft (the glomerulus) into the surrounding protrusion (Bowman's capsule), across which it flows into the lumen of the long tubular system that drains into the urinary bladder and ultimately out of the body.

From this ultrafiltrate (which is already functionally *outside* of the body) all the ions and small solutes that the body needs must be reabsorbed back into the body. The loss of these ions and small solutes from the body would constitute an enormous waste of the metabolic energy invested in their initial synthesis or absorption from the intestine. This reabsorption is done by a series of complex, integrated processes as the

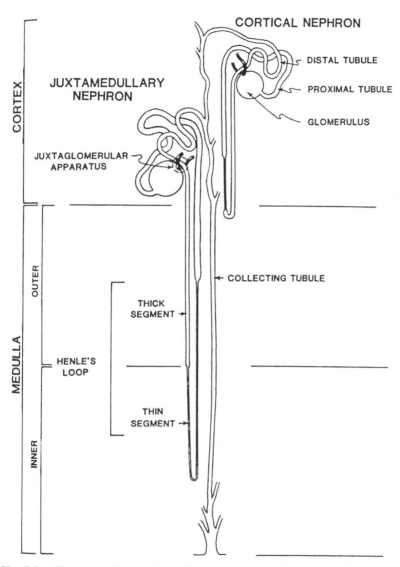

Fig. 7.8.   The mammalian nephron. Two nephrons, one of the cortical class and the other of the juxtamedullary class, connected to a single collecting tubule. In this section, we will be particularly concerned with the proximal tubule, the thick segment (or thick ascending limb), and the collecting tubule or duct. Taken, with kind permission, from L. P. Sullivan and J. J. Grantham (1982). "Physiology of the Kidney" 2nd Edition, Philadelphia, Lea and Febiger.

urine flows down the tubules of the nephron. In addition, there are a few specific points along the tubular system at which occurs the actual secretion of certain particular substances that the body gets rid of by this means.

For our purposes, we can divide this long tubular system into two sections: the *proximal* (or nearest) *tubule* into which the urine first flows after it drains into Bowman's capsule and the *distal* (or furthest) *tubule,* which connects the proximal tubule to the *collecting duct* which itself drains into the bladder. The distal tubule is composed of three parts, only one of which, the *thick ascending limb* (TAL), will we find it necessary to consider. Each section of the nephron has a particular role in ensuring the efficient reabsorption of the needed materials from the urine. The nephron is surrounded by capillaries bringing in a rich blood supply. These drain the *interstitial fluid* (that between the cells), which bathes the basolateral surfaces of the tubule cells. The apical cells face the lumen of the tubule and, hence, the urinary fluid that the kidney is engaged in processing.

The reader will note that each of the three epithelial types is found in a different grouping in Table 7.1. The proximal tubule is a "leaky" epithelium (and, from our earlier discussion likely to be involved in bulk, but selective, fluid movements), the thick ascending limb is of the "intermediate" type (and is concerned in separating sodium chloride from water), while the collecting duct is a "tight" epithelium and functions as an effective barrier to ions and water.

7.2.1.4. THE TRANSPORT SYSTEM "MENU"

It can be helpful to consider the various epithelial tissues as if each were a two-course dinner made up from a "menu" of different possible "dishes" for each course. The "first course" is chosen from among the range of transport systems that can be present at the apical surface [depicted in Fig. 7.9(a)]. The "second course" is chosen from among the transport systems depicted in Fig. 7.9(b), showing the range that is available at the baso-lateral surface. Each surface can have channels (depicted as pores) selective for different ions, and/or co- or countertransport systems (symports or antiports). The sodium–potassium pump is always present at the basolateral surface and only at that surface, and this surface can also contain facilitated diffusion (uniport) systems for various substrates. Depending on the precise arrangement of the various transport systems that it possesses, an epithelium can be involved in the *absorption* of a substrate from the fluid bathing the apical surface to deliver it at the basolateral surface, or in its *secretion* in the opposite direction.

Consider now the three kidney epithelial systems that we have chosen to discuss.

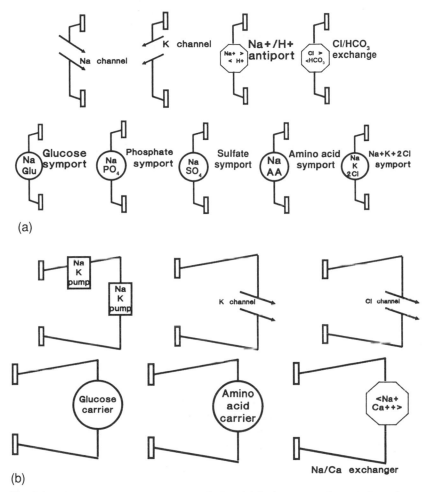

Fig. 7.9. The transport systems "menu". Part (a) depicts many of the choices of transport systems that are available at the apical face of an epithelium; (b) the choices at the basolateral face. Channels are represented as parallel arrows, carriers (and cotransporters) as circles, antiporters as octagons, and pumps as squares.

7.2.2. A Tight Epithelium: The Collecting Duct

Easiest to understand is the tight epithelium, the collecting duct. This has, at its apical surface, a highly selective sodium channel and at its basolateral surface a potassium channel and, of course, the sodium–potassium pump (Fig. 7.10).

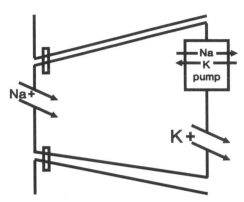

Fig. 7.10.   A tight epithelium. The apical face of the cell (left) contains a sodium-selective channel; the basolateral face, the sodium–potassium pump, and a potassium–selective channel.

The cells of this epithelium are in close apposition, side by side, the whole forming a sheet that is very tight. Consider what happens when this tissue is functioning. The sodium pump, using ATP as the provider of energy, pumps out sodium ions from the cell interior across the basolateral surface, from where they are carried out into the interstitial fluid that bathes this surface. Ion flow down basal and apical gradients leads to an electrical voltage being built up across the epithelium, directed so that the medium bathing the basolateral surface is at a more positive potential than that bathing the apical surface. The depletion of *cellular sodium* that results from the action of the pump intensifies the electrochemical gradient of sodium between the lumen of the duct and the cell interior, so that sodium flows into the cell along the sodium channel. There is thus a net movement of sodium from the lumen of the duct, back into the blood. [This is a reabsorption since the sodium ions originally came from the blood, from where they were filtered across the glomerulus (Fig. 7.8).] In contrast, potassium ions that enter the cell during the action of the sodium (+ potassium) pump, circulate out again to the blood along the potassium channel that is present in the basolateral face of the cell. Chloride ions are driven across the epithelium by the transmembrane potential developed by the flow of the cations. The chloride may move through a special subpopulation of the epithelial cells. The epithelium, being tight to water movement, thus allows the collecting duct to abstract sodium chloride selectively from the lumen, leaving a more dilute urine to flow into the urinary bladder.

A very similar arrangement of transport systems functions in other tight epithelia, such as urinary bladder, colon, and amphibian skin, all tissues that functionally can transport sodium chloride, but very little water from the apical to the basolateral surface. In the colon and in amphibian skin, such a system is of utmost importance in the nutritive uptake of sodium chloride.

When an animal becomes dehydrated, special rehydrating systems are brought into operation. Short-term regulation results from the action of a specific hormone, the antidiuretic hormone (ADH), also known as *vaso-pressin*. ("Diuresis" is the clinical term for an excessive flow of urine.) ADH is released into the blood of the dehydrated animal and reaches the *basolateral* face of the tight epithelium of the kidney. There it reacts with its specific receptor, setting off a train of events that lead to an increase in the level of a "second messenger," cyclic AMP (see any textbook on biochemistry) within the cell. Very rapidly, a great increase in the water permeability of the *apical* membrane ensues. What was a very water-impermeable membrane now becomes highly permeable and water flows back into the bloodstream and hence into the body instead of into the urinary bladder.

Precisely the same train of events, following the same signal, occurs in amphibian skin, and water flows across the skin from the pond to rehy-drate the animal. An additional switching event occurs in amphibian skin: the sodium channel at the apical face is stimulated so that more sodium flows into the cell and out at the basolateral surface, where it increases the osmotic concentration there, encouraging the flow of water across the skin and into the animal. In rat and mouse kidney, however, the increase in sodium flow occurs in another part of the organ, the thick ascending limb, as we shall discuss in the next section. The net effect is the same in that the sodium chloride concentration at the basolateral surface of the collecting duct is increased, amplifying the osmotic gradient there and hence the flow of water out of the lumen, back into the body.

One more specialization is found in certain cells of the collecting duct. This is a *potassium-selective* channel at the apical face of these particular cells. Such a channel allows potassium ions to flow from the cell interior, where its concentration is kept high by the sodium–potassium pump, into the lumen of the collecting duct and hence out into the urine to be stored in the bladder. Thus potassium can be *secreted* out of the body. These potassium-selective channels are switched on when body potassium levels are unfavorably high for the animal. This is another case of the regulation of transport activity.

We saw that the water permeability of the tight epithelia increases dramatically when the animal is dehydrated and the hormone ADH is released into the bloodstream. In the collecting duct of the rabbit, the change in apical (osmotic) water permeability is almost fivefold, while the

Box 7.3. Sodium and Water Channels of Tight Epithelia

We should take a closer look at two of the specialized transport systems of these tight epithelia: their sodium channels and their water channels. A characteristic feature of the sodium channels of tight epithelia is that they are *very* selective for sodium ions. Indeed, of the many ions that have been tested, only lithium and hydrogen ions seem to be able to pass these channels to any great extent. In the epithelium of the collecting duct the channels pass sodium ions 10 times more readily than they pass potassium ions, but lithium ions permeate 50% more rapidly than do sodium ions. The single channel conductance for sodium is a low 5 pS (compare Table 3.2). When an attempt was made to force sodium ions through such a channel by applying an osmotic gradient across a urinary bladder epithelium (refer to the discussion on electroosmosis in Box 2.5), no sodium current could be measured, suggesting that water does not move through the sodium channel. The channel seems to behave as a size-selective filter discriminating against any particle bigger than a naked sodium ion. Another characteristic of these sodium channels is that they are *regulated* by the concentration of sodium ions that bathe them. As the sodium ion concentration at the apical surface is raised, the number of open sodium channels decreases and the net flow of sodium across the epithelium is reduced. A similar regulatory decrease of channel activity takes place as the sodium concentration *within* the epithelial cells is raised. Thus the sodium-absorbing capacity of the epithelium is maximal when the animal's need for sodium uptake is greatest, that is, when it has low intracellular sodium and is in a low-sodium environment. These sodium channels are inhibited by the drug amiloride. Amiloride is itself positively charged and might be thought to bind at sites on the channel at which sodium would otherwise bind. These sites may be the regulatory, rather than the transport, sites for sodium.

The water channels at the apical surface of the tight epithelia have been intensively studied. As we saw in Section 2.9, measurements of water flow under an osmotic gradient, when compared with its rate by diffusion, can be used to estimate the length of a channel if it is long and thin, or its width if it is wide. As the water channel at the apical face of the tight epithelia does not admit urea, a small molecule, the channel appears to narrow at some point. The presence of unstirred layers of water on the surfaces of such epithelia (and within the cells

(Continued)

that compromise them) introduces a difficulty in the interpretation of such data. When, however, the contribution of these unstirred layers is properly taken into account, osmotic flow often remains an order of magnitude greater than diffusive flow, when measured in the same units (Section 2.9). It appears that at least 10 water molecules might well be able to lie in line along the length of such water channels. The water channels allow the passage of protons but not of sodium or potassium ions.

water permeability of the basolateral membranes does not increase to any significant extent. The change in water permeability takes place very rapidly; in the urinary bladder of the toad it is completed within 30 min. This is far too short a time to allow the synthesis within the cell of new protein channel material. Rather, it appears that there is a store of channel material in the epithelial cells, located in vesicles termed *aggrephores*. These fuse with the apical membranes when the cell is stimulated by ADH, donating their water channels to the apical membranes. The time course of aggrephore fusion with the apical membrane parallels the time course of the increase in water permeability. A number of proteins have been isolated from the aggrephores and information as to whether these proteins can, indeed, act as water channels is eagerly awaited.

The hormone aldosterone has a profound long-term effect on stimulating sodium movement across tight epithelia. Its effect, in contrast to that of ADH, requires the synthesis of new proteins and hence several hours elapse between administration of the hormone and the full augmentation of sodium transport. The synthesis of new sodium pumps has been directly demonstrated. Both α and β chains are synthesized and can be seen to be inserted into the membrane of the basolateral surface.

7.2.3. An "Intermediate" Epithelium: The Thick Ascending Limb of the Mammalian Kidney

The thick ascending limb (TAL) of mammalian kidney is richly endowed with sodium–potassium pumps and, in rodents, receptors for the hormone ADH. It is a section of the nephron that is well equipped for its role as part of the water-conserving mechanism, which it carries out indirectly, as we shall see. In terms of its overall resistance, it is intermediate between the tight and very leaky epithelia. Let us consider the "menu" of Fig. 7.9 in its application to the TAL. Its sodium pump is, of

course, situated at the basolateral surface. At the apical surface is the
$Na^+ + K^+ + 2Cl^-$ cotransporter, which we discussed in Section 5.3.6.
(Recall that this cotransporter is inhibited by the diuretic agents furosemide and bumetanide, emphasizing its role in water movements within
the kidney.) Consider what happens when these two systems are in operation, arranged in the vectorial fashion shown in Fig. 7.11. The sodium
pump reduces sodium levels in the cell to far below those to which they
would reach in the absence of a metabolic energy input. This electrochemical gradient of sodium is then harnessed by the $Na^+ + K^+ + 2Cl^-$
cotransporter, which drives chloride ions into the cell from the apical
surface, building up an electrochemical gradient of this chloride. The
chloride ions now flow out of the cell at the basolateral surface along a
specific chloride channel located there. The net result so far is a flow of
two chloride ions across the epithelium for each sodium ion that is
pumped out at the basolateral face. This excess flow of the chloride ion
develops an electrical potential *across* the epithelium, negative in the
solution bathing the basolateral face. Since the epithelium of the TAL is
somewhat leaky, sodium ions can flow *between* the cells, driven by the
electrical potential, adding to the overall net flow of sodium. The sum of
these effects is that sodium chloride is driven by two routes across the
epithelium, and indeed twice as much is driven by the combined action of
the transport systems as by the sodium pump alone. (We still must account for the potassium movements. Potassium enters the cell from the
basolateral face on the sodium pump and leaves in part at that face across
a potassium channel. It also enters at the apical surface on the cotransporter and recycles back at that face on a channel also situated there. There

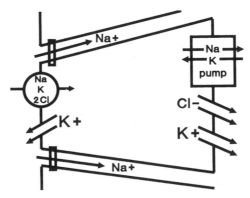

Fig. 7.11.   An intermediate epithelium. The apical face (left) contains a potassium channel and a sodium–chloride cotransporter. The basolateral face possesses the sodium–potassium pump, a chloride and a potassium channel.

remains a small net secretion of potassium from blood to the lumen of the TAL.)

The major overall effect of these transport systems in the TAL is then that sodium chloride is drawn out of the fluid within the tubule, to be reabsorbed by the body. The fluid within the tubule thus becomes *diluted* as it passes through the TAL, and this segment of the kidney is, therefore, sometimes called the diluting segment. It possesses this property since it is of only intermediate leakiness. Its water permeability is relatively low, and thus an osmotic gradient can develop across the epithelium of the TAL. As its ion permeability, however, is relatively high, a chloride-dominated transepithelial potential can be harnessed to provide an energy-sparing, additional transport of sodium chloride.

What are the effects of the antidiuretic hormone, released into the bloodstream when a rat or mouse is suffering dehydration stress? ADH binds to its receptors situated on the exterior (blood-facing) surface of the basolateral face. These receptors are activated, setting off a train of events that lead to a rise in the concentration of cyclic AMP in the cell. This, in turn, leads to the *phosphorylation* of a protein linked to the $Na^+ + K^+ + 2Cl^-$ cotransporters, increasing their activity and driving more sodium and chloride ions (and potassium) into the cell. The sodium pump is then able to pump more sodium out of the cell at the basolateral face. Chloride, coming into the cell on the cotransporter at the apical face, flows out at the basolateral face, along the channels situated there, developing a transepithelial potential and pulling with it more sodium. This finally leads to an overall net increase in the flow of sodium chloride across the epithelium. The salt is deposited in the interstitial fluid of the kidney, in the region around the TAL. A high salt concentration builds up around the TAL, and some of this salt then diffuses along its concentration gradient within the interstitial tissue, outside of the tubular structures. When the region around the collecting ducts becomes rich in salt, and the water permeability of the collecting duct is increased markedly by the action of ADH (see Section 7.2.2), water is encouraged to flow out of the tubular system and back into the body, conserving the water resources of the animal. We now see why inhibitors of the $Na^+ + K^+ + 2Cl^-$ cotransporter act as diuretics (i.e., increase water flow out of the body).

7.2.4. A Leaky Epithelium: The Proximal Tubule

The proximal tubule is the most complex of the kidney's segments, from a transport point of view. Indeed, it is best considered as a series of differently specialized segments rather than as a uniform whole, with each segment displaying a different set of transport systems chosen from our "menu" of such systems. Consider the first segment, the so-called proximal convoluted tubule, in which the transport systems are as depicted in Fig. 7.12(a). As usual, we have the sodium pump situated at the baso-lateral face, ensuring that the electrochemical gradient of sodium is everywhere such that there is a tendency for sodium ions to diffuse back into the cell. At the apical face is an array of sodium-dependent cotransport systems. Different systems carry glucose, amino acids, carboxylic acids,

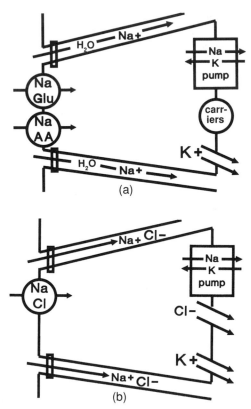

Fig. 7.12. Two leaky epithelia. (a) A diagram of the situation in the first part of the proximal tubule. The apical membrane (left) contains cotransporters for sugars (Glu) and amino acids (AA). The basolateral membrane contains the sodium pump, a potassium channel, and carriers for the sugars and amino acids. Sodium and water move between the cells. (b) A diagram for the latter part of the proximal tubule. A system that has the effect of cotransporting sodium and chloride ions (see text) is at the apical face. At the basolateral face is the sodium pump and a chloride and potassium channel. Sodium chloride moves in great part between the cells.

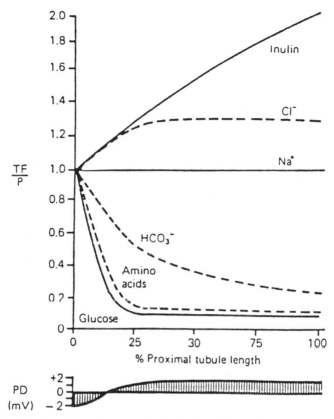

Fig. 7.13. Reabsorption of water and solutes along the proximal tubule. Concentration ratio of the various solutes in the tubular fluid as compared with their concentration in the plasma is symbolized by *TF/P*. A value greater than unity means that the substance in question is concentrated within the tubule; a value less than one, that it has been pumped out of the tubule, back into the body. Inulin cannot cross the tubular epithelium, and hence its concentration within the tubule is a measure of how much water has been pumped out of the tubule. Taken, with kind permission, from F. C. Rector (1983). *Am. J. Physiol.* **244,** F461–471.

phosphate, and sulfate out of the lumen of the tubule and into the cell. These critically important metabolites are those that were filtered across the glomerulus (see Section 7.2.1.1) from the blood into the first-formed urine. The transport systems of the proximal tubule are responsible for their reabsorption, preventing their loss into the urine. Within the first quarter of the distance along the proximal tubule, all of these nutrients are largely reclaimed from the urinary fluid (see Fig. 7.13).

How do these metabolites, which are all of the type of water-soluble molecule that cross cell membranes only with difficulty, escape across the basolateral face of the tubule cell? The answer is that specialized facilitated diffusion systems—carriers—for each of these metabolites are present at the basolateral face. The glucose system, for instance, is one closely related to the facilitated diffusion system of the red blood cell that we considered in Sections 4.1 and 4.2. The sodium-dependent cotransporter of glucose at the apical face of the tubule harnesses the inwardly directed electrochemical gradient of sodium to drive glucose into the tubule cell. The glucose carrier at the basolateral face enables this glucose to leave the cell, by facilitated diffusion down its chemical gradient from the tubule cell into the bloodstream. A similar arrangement of cotransporters and carriers at apical and basolateral faces brings the amino acids and inorganic ions back into the blood from the urinary fluid. Sodium bicarbonate appears to be reabsorbed by a different mechanism (which we have not depicted in Fig. 7.12). The *apical* face of the tubule cell possesses the antiporter for sodium and protons that we discussed at the beginning of the present chapter, the electrochemical gradient for sodium being harnessed to drive protons *out* of the cell and into the urinary fluid. In addition, the membrane at this apical surface possesses an exchanger of bicarbonate and chloride, the anion carrier that we discussed in Section 4.8.1. The combined operation of the two exchangers would result in the uptake of sodium bicarbonate into the tubule cell in exchange for the secretion of hydrogen chloride into the lumen. The secretion of acid is of great importance for the maintenance of the acid–base balance of the body and represents another, major transport function of the kidney.

As a result of the activity of all these sodium-coupled transporters at the apical face of the cell and of the sodium pump and the specialized carriers at the basolateral face, there is a substantial overall flow of sodium and accompanying metabolites across the tubule from urinary fluid to blood. The tubular epithelium here is very water-permeable so this flow of sodium and metabolites drives an *osmotic* flow of water from the urinary fluid to the blood. Most of the water lost into the kidney is reabsorbed into the body along the length of the proximal tubule as a result of this osmotic flow, driven primarily by the pumping of sodium. Figure 7.13 shows how the urinary fluid loses water as it flows along the proximal tubule. The quantity plotted on the y-axis, *TF/P*, is the ratio of the concentration of a particular substance in the tubular fluid compared to its concentration in the plasma. A value for this ratio greater than unity means that the substance in question is being *concentrated* in the urine at that point. "Inulin," on the uppermost curve, is a water-soluble polymer that enters the urine by being filtered across the glomerulus but is not transported by any of the transport systems in the nephron. It is concentrated in the urine as a result of water moving back into the body. Its concentration in the tubular fluid is thus a measure of the absorption of water out of the urine. The concentrations of the remaining solutes generally drop despite the abstraction of this water from the fluid. Note, too, that the concentration of chloride ions initially *increases* as the water is reabsorbed back into the blood. There are few if any transport routes for chloride in the early part of the proximal tubule, a fact that is of great importance for the overall economy of the tubule's activity, as we will now discuss.

Consider what happens when the urinary fluid reaches the latter portion of the proximal tubule, the so-called proximal straight segment. By now, there has been almost complete removal of metabolites such as the sugars and amino acids, and the removal, too, of much of the bicarbonate. The chloride ions that remain are now at a *higher* concentration within the tubule than they are present in the blood. In this segment of the proximal tubule, chloride ions are enabled finally to diffuse across the epithelium, down their concentration gradient, through a chloride-conducting pathway, which seems to be situated *between* the cells. Diffusion of this chloride sets up a diffusion potential (Section 3.5), ensuring that the lumen of the tubule becomes *positive* with respect to the surface bathed by the blood. This positive potential drives sodium ions across this leaky epthelium, *between* the cells of the tubule [see Fig. 7.12(b)]. The overall effect is that additional sodium chloride moves across the epithelium and is reabsorbed back into the body. Note one very important fact: this movement of sodium chloride is not a direct result of sodium pumping but rather an indirect one. Because the movement of chloride is delayed until the latter portion of the proximal tubule, it moves *down* its electrochemical gradient in an energy-sparing fashion. The net effect is that some twice as much sodium is pumped across the proximal tubule for every ATP split than is accounted for by the sodium pump working alone! Clearly this energy-sparing can be achieved only because the epithelium of the proximal tubule is leaky, allowing the movement of sodium chloride by a route between the cells rather than across them.

In addition to this movement of chloride *between* the cells, there is a co-transport of sodium + chloride across the apical face of the cells of the proximal tubule [see Fig. 7.12(b)]. There is good evidence that this is not a simple symport but rather a coupling of two antiporters similar to the scheme depicted in Fig. 7.4. In the case of the proximal tubule, however, one antiporter is indeed the Na^+/H^+ antiporter but the second seems to exchange chloride and *formate* ions across the membrane.

7.2.5. Tight, Intermediate, and Leaky Epithelia Compared

It must not be thought that these three types of epithelia are present only in kidney. Table 7.1 lists a range of tissue types for each class of epithelium. For each tissue, however, certain key physiologic functions determine which class of epithelium is selected, according to its transport characteristics, as follows:

1. Tight epithelia are found where water movement is to be minimal or else very rigorously controlled, while the pumping of sodium chloride has to take place against a steep concentration gradient. It would appear that all of the pumped sodium here is pumped with the direct harnessing of metabolic energy by the sodium-dependent ATPase. Sodium entry across the apical face is not steeply downhill and is not harnessed to the flow of any other solute, but enters rather through a highly selective channel.

2. An intermediate epithelium, in contrast, moves salt up a small concentration gradient, but moves it unaccompanied by a flow of water, building up in the process an osmotic gradient. Since the salt flow is not up a steep gradient, it can be performed in an energy-sparing fashion and sodium transport driven directly by the pump is coupled to the transport of additional sodium, through the paracellular route.

3. Leaky epithelia are found where vast quantities of water are required to move with no change in the osmotic concentration of the fluid transported. Sodium chloride is moved up a very small overall gradient, if any, and the direct use of ATP energy by the sodium pump is amplified effectively by a set of coupled transport systems. Refer to Table 5.2 which lists how various ions and metabolites are transported across a number of different epithelia and demonstrates the range of transport systems that can be found.

Box 7.4. Transcellular Transport of Calcium

A Special Case: The Ferry-Boat System

Calcium is a very important bodily constituent and appropriate epithelial transport systems exist in order to extract calcium from the diet or to reabsorb it from the urinary fluid. The fact that calcium is present in such low concentrations in most cells (see Sections 6.2 and 6.3)

(Continued)

presents, however, a special difficulty. If the "menu" for the active transport of calcium across an epithelium were to consist only of a calcium pump at the basolateral face and a calcium-admitting pathway at the apical face, this would leave calcium movement *across* the cell to be driven by the gradient of free calcium.

Consider an intestinal cell, about 20 μm in length, from a rat. The concentration gradient of free calcium across the cell from apical to basolateral face is probably only about 100 nM. We can calculate the maximal diffusional flow of calcium across the cell that this gradient can support if we substitute the appropriate values into the diffusion equation [Eq. (2.2) in Section 2.1]. Taking a value for the diffusion coefficient of calcium in water at 37°C as 2.1×10^{-5} cm^2/sec, we calculate that the maximal diffusional flow of calcium across the cell is 1.6×10^{-18} mol/sec per cell. The *measured* maximum flux in a calcium deprived rat whose calcium uptake system is regulated to maximum ability is 1.2×10^{-16} mol/sec per cell, over 70 times more than this "maximal possible" value. For the rat kidney, one can show similarly that the measured maximal flux is 80 times the flux calculated assuming that it is the gradient of *free calcium* that supports the flow of calcium across the cell. How is this apparant paradox to be resolved? It appears that the epithelia concerned in calcium absorption possess a specialised protein, a "calcium-binding" protein (calbindin) that is present in such cells in *millimolar* concentrations. At all points along the epithelial cell, free calcium, and the calcium-binding protein are in equilibrium. Where the level of free calcium is low, less calcium is bound to calbindin, where free calcium is high, more is bound (see Fig. 7.14).

Thus the small transcellular concentration gradient of free calcium is amplified into a large concentration gradient of bound calcium. It is this gradient that supports the transcellular flux of calcium, with the calcium-binding protein ferrying calcium across the cell. If the appropriate substitutions into the diffusion equation [Eq. (2.2) in Section 2.1) are made, it can be readily shown that realistic values for the transcellular flux of calcium can now be calculated. The ability of an epithelium to transport calcium is *regulated* by the concentration of calbindin. Vitamin D is a hormone that controls the body's ability to absorb calcium from the intestine and to reabsorb it in the kidney. Vitamin D serves as a signal controlling the *expression* of calbindin in those tissues where it is needed for calcium transport. If the body is deprived of calcium, vitamin D levels rise and the synthesis of calbindin is stimu-

(*Continued*)

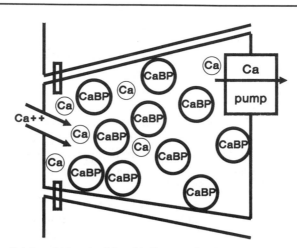

Fig. 7.14.   Calcium (Ca) and calcium-binding protein (CaBP) in equilibrium across an epithelium. Calcium enters the eptihelial cell via a channel on the left and is removed by a pump on the right. Across the cell is a gradient of calcium ions and an amplified gradient of CaBP.

lated in intestine and kidney. Calcium can be ferried rapidly across the cell and pumped into the body. If calcium levels are high, vitamin D levels fall, the synthesis of calbindin is reduced and calcium is lost from the body.

At the basolateral face of the appropriate epithelia, less striking effects of vitamin D can be seen on the expression of the calcium pumping systems. These systems comprise an ATP-driven calcium pump (as we discussed in Section 6.3) and a sodium–calcium exchanger (see Section 5.2.5). The system that admits calcium at the apical face of the cell seems to be a calcium channel.

7.2.6. Vectorial Assembly and Sorting of Membrane Transport Systems in Epithelia

Effective functioning of epithelia requires that they be organized in a vectorial, polar fashion. Apical and basolateral surfaces of the epithelial cells have quite different protein constituents. These proteins are subject to regulation and, in many cases, proteins are newly synthesized in response to hormonal signals and are conveyed accurately to the appropriate membrane, apical or basolateral. In addition, such membrane proteins are being continuously *recycled*. Proteins are removed from the plasma membranes and carried into the cell interior in endocytotic vesicles, whereupon many of them return to that membrane (either apical or basolateral) from which they were removed. This membrane traffic is thus fully regulated. We do not yet know how this is achieved, what the signaling systems involved are, or how a particular protein is recognized as being apical or basolateral and appropriately funneled to one or the other membrane. It seems, however, that the tight junctions do form a barrier

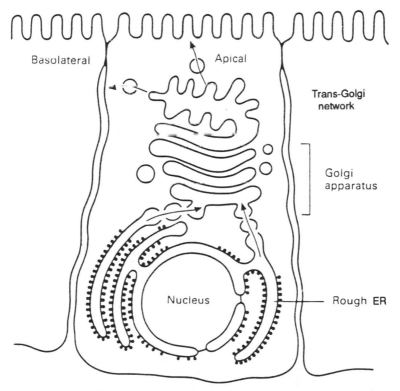

Fig. 7.15.   Scheme for the sorting of apical and basolateral proteins. The proteins are synthesized in the endoplasmic reticulum (Rough ER), transported and modified through the Golgi apparatus and finally sorted from each other in the trans-Golgi network. The vesicles that reach the apical or basolateral membranes fuse there, delivering their cargo locally. Reprinted from K. Simons (1987). *Kidney International* **32** Suppl 23, S201–S207, with permission.

that prevents the mingling of the membrane proteins from the two halves of the cell. Figure 7.15 depicts protein synthesis in an epithelial cell and the flow of proteins to apical and basolateral membranes. Labeling experiments show that the dispatch of proteins to one or other surface takes place after they have left the so-called trans-Golgi network, the last stage of protein synthesis and processing (see Section 1.5 and any cell biology textbook).

We saw in previous sections of this chapter that the tight junctions can form a barrier to the passage of water and solutes *across the epithelium* from luminal or mucosal to serosal surface. The tight junctions form a barrier, too, to the mixing *in the plane of the membrane* of some, but not all, of the lipid constituents of the cell (see Fig. 7.16). When the apical membranes of kidney cells (studied in cell culture) were labeled with a phospholipid that marked the *outward facing half* of the membrane bilayer, no label entered the basolateral membrane. When the *cytoplasmic facing half* of the apical membrane was labeled, however, the label entered the basolateral membranes freely. Thus the tight junctions can act as a fence, preventing the lateral diffusion of some of the phospholipid constituents of the membranes. It is thus not surprising that these junc-

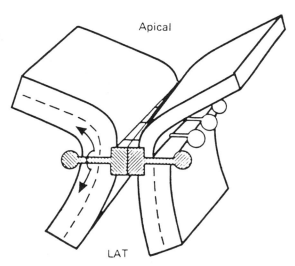

Fig. 7.16. A schematic model of the tight junction. The diagram shows how protein molecules embedded within the apical membranes of one cell may interact tightly with those in the adjacent cell to form the tight junction. The arrow shows that in the inner half of the lipid bilayer, these lipid molecules diffuse freely, whereas in the outer half diffusion of lipids and, of course, the proteins is limited. Reprinted from K. Simons (1987). *Kidney International* **32** Suppl 23, S201–S207, with permission.

tions can separate effectively the apical and basolateral proteins and determine the polar nature of the epithelial cells.

Clearly, much remains to be learned about the functioning and role of the epithelial cells. They represent the organization, regulation, and integration of membrane transport systems at the highest level. Their study has provided much insight into bodily function and will provide even more as research advances.

SUGGESTED READINGS

General

Mandel, L. J., and Benos, D. J., eds. (1986). "The Role of Membranes in Cell Growth and Differentiation," Curr. Top. Membr. Transp., Vol. 27. Academic Press, Orlando, Florida.

Cell Volume: Long Term

Jakobsson, E. (1980). Interactions of cell volume, membrane potential and membrane transport parameters. Am. J. Physiol. 238 (Cell Physiol. 7), C196–C206.
MacKnight, A. D. C. (1987). Volume maintenance in isosmotic conditions. Curr. Top. Membr. Transp. 30, 3–43.
MacKnight, A. D. C., and Leaf, A. (1977). Regulation of cell volume. Physiol. Rev. 57, 510–573.
Post, R. L., and Jolly. P. C. (1957). The linkage of sodium, potassium, and ammonium active transport across the human erythrocyte membrane. Biochim. Biophys. Acta 25, 118–128.

Cell Volume: Short Term

Cala, M. (1986). Volume-sensitive ion fluxes in Amphiuma red blood cells: General principles governing Na-H and K-H exchange transport and Cl-HCO₃ exchange coupling. Curr. Top. Membr. Transp. 27, 194–218.
Grinstein, S., et al. (1986). Activation of the Na^+-H^+ antiport by changes in cell volume and by phorbol esters; possible role of protein kinase. Curr. Top. Membr. Transp. 26, 115–134.
Haas, M., and McManus, T. J. (1985). Effect of norepinephrine on swelling-induced potassium transport in duck red cells: Evidence against a volume-regulatory decrease under physiological conditions. J. Gen. Physiol. 85, 649–667.
Hoffman, E. K. (1987). Volume regulation in cultured cells. Curr. Top. Membr. Transp. 30, 125–180.
Spring, K. R. (1985). Determinants of epithelial cell volume. Fed. Proc., Fed. Am. Soc. Exp. Biol. 44, 2526–2529.

Kidney Structure

Jamison, R. L., and Kriz, W. (1982). "Urinary Concentrating Mechanism: Structure and Function." Oxford Univ. Press, New York.

Kriz, W., and Kaissling, B. (1985). Structural organization of the mammalian kidney. *In* "The Kidney: Physiology and Pathophysiology" (D. W. Seldin and G. Giebisch, eds.), pp. 265–306. Raven Press, New York.

Types of Epithelia

Diamond, J. M. (1978). Channels in epithelial cell membranes and junctions. *Fed. Proc., Fed. Am. Soc. Exp. Biol.* **37**, 2639–2644.
Frömter, E., and Diamond, J. M. (1972). Route of passive ion permeation in epithelia. *Nature (London), New Biol.* **235**, 9–13.
Macknight, A. D. C., *et al.* (1980). Sodium transport across toad urinary bladder: A model "tight" epithelium. *Physiol. Rev.* **60**, 615–715.

Transport System Menu

Lauf, P. *et al.* (1987). Physiology and biophysics of chloride and cation cotransport across cell membranes. *Fed. Proc., Fed. Am. Soc. Exp. Biol.* **46**, 2377–2394.
Molony, D. A., *et al.* (1989). Na$^+$:K$^+$:2Cl$^-$ cotransport and the thick ascending limb. *Kidney Int.* **36**, 418–426.
Reuss, L. (1989). Ion transport across gallbladder epithelium. *Physiol. Rev.* **69**, 503–545.
Sacktor, B. (1989). Sodium-coupled hexose transport. *Kidney Int.* **36**, 342–350.
Zelikovic, I., and Chesney, R. W. (1989). Sodium-coupled amino acid transport in renal tubule. *Kidney Int.* **36**, 351–359.

Water Channels

Abramow, M. *et al.* (1987). Cellular events in vasopressin action. *Kidney Int.* **32**, Suppl. 21, S-56–S-66.
Hays, R. M. *et al.* (1987). Morphological aspects of the action of ADH. *Kidney Int.* **32**, Suppl. 21, S-51–S-55.
Jamison, R. J. (1987). The renal concentrating mechanism. *Kidney Int.* **32**, Suppl. 21, S-43–S-50.
Schafer, J. A. (1984). Mechanisms coupling the absorption of solutes and water in the proximal nephron. *Kidney Int.* **25**, 708–716.

Calcium Transport across Epithelia

Bronner, F. *et al.* (1986). An analysis of intestinal calcium transport across the rat intestine. *Am J. Physiol.* **250** (*Gastrointest. Liver Physiol.* **13**), G561–G569.

Vectorial Assembly and Sorting

Alberts, B. *et al.* (1989). "The Molecular Biology of the Cell," 2nd ed., Chapter 6. Garland Publishing, New York.
Simons, K. (1987). Membrane traffic in an epithelial cell line derived from the dog kidney. *Kidney Int.* **32**, Suppl. 23, S-201–S-207.

APPENDIX 1

Single- and Triple-Letter Codes for the Amino Acids

Amino acid	Single-letter code	Triple-letter code
Alanine	A	Ala
Cysteine	C	Cys
Aspartic acid	D	Asp
Glutamic acid	E	Glu
Phenylalanine	F	Phe
Glycine	G	Gly
Histidine	H	His
Isoleucine	I	Ile
Lysine	K	Lys
Leucine	L	Leu
Methionine	M	Met
Asparagine	N	Asn
Proline	P	Pro
Glutamine	Q	Gln
Arginine	R	Arg
Serine	S	Ser
Threonine	T	Thr
Valine	V	Val
Tryptophan	W	Trp
Tyrosine	Y	Tyr

APPENDIX 2

Fundamental Constants, Conversion Factors, and Some Useful Approximations

Fundamental Constants

R (gas constant)   8.314 J(joules) deg^{-1} mol^{-1}
1.987 cal deg^{-1} mol^{-1}

F (Faraday)   96,500 J V^{-1} (volt)$^{-1}$ equiv^{-1}
(or coulombs equiv^{-1})
23,060 cal V^{-1} equiv^{-1}

N (Avogadro number) 6.02×10^{23} particles mol^{-1}

e (electronic charge) 1.602×10^{-19} coulombs ($= F/N$)

Conversion Factors

$0°C = 273$ K

1 cal = 4.184 J

1 electron volt (eV) = 23.06 kcal/mol = 96.5 kJ/mol

(Thus, to convert mV to kJ/mol, multiply by 0.0965; to convert to kcal/mol, multiply by 0.0231.)

$2.3RT/F = 59$ mV at 25°C

(Thus, a 10-fold ratio of concentrations at 25°C is equivalent to 59 mV or 1.36 kcal/mol or 5.69 kJ/mol for a univalent ion.)

Some Useful Approximations (for back-of-the-envelope calculations)

Thickness of cell membrane permeability barrier	$= 5$ nm
Capacity of cell membrane	$= 1$ μF/cm^2
Dielectric constant of aqueous fluids	$= 80$
Dielectric constant of membrane interior (lipids)	$= 2\text{--}4$
Radius of "typical" animal cell	$= 20$ μm
Radius of human red blood cell	$= 5$ μm
Surface area of "typical" cell	$= 5000$ μm^2
Surface area of human red blood cell	$= 140$ μm^2
Volume of "typical" cell	$= 35{,}000$ μm^3 (35 pl)
Volume of human red blood cell	$= 90$ μm^3 (0.09 pl)
Typical transmembrane potentials	$= -60$ to -80 mV (See Table 2.2.)
Aqueous diffusion coefficients of metabolites	$= 0.5\text{--}2 \times 10^{-5}$ cm^2 sec^{-1} (See Table 2.1.)
Permeability coefficients of small hydrophilic permeants across plasma membranes	$= 10^{-9}$ to 10^{-3} cm sec^{-1} (See Table 2.3.)
Single channel conductances	$= 4\text{--}200$ pS (See Table 3.2.)
Turnover numbers of carriers and pumps	$= 10\text{--}2000$ sec^{-1} (See Table 4.2.)

APPENDIX 3

Relationship Between Permeability Coefficient P_s and Half-Time $t_{1/2}$ of Entry of a Permeant

Take the fundamental Eq. (2.7) in Section 2.3:

$$J_{I \to II} = P_s A(S_I - S_{II}) \qquad (2.7)$$

where $J_{I \to II}$ is the flux of the substrate S from side I to II across surface area A of the membrane for a substance of permeability coefficient P_s. Now rewrite this in terms of the number dn of molecules of the substrate entering a cell of volume V during time dt:
Then

$$dn/dt = PA(S_I - n/V) \qquad (A3.1)$$

where S_I is the constant concentration of S outside the cell, while n/V is the concentration inside when n molecules have entered.
Separating the variables, we have

$$dn/(VS_I - n) = PA \, dt/V \qquad (A3.2)$$

Integrating both sides of this equation, we obtain

$$-\ln(VS_I - n) \Big]_0^n = PAt/V \Big]_0^t \qquad (A3.3)$$

At time zero, $t = 0$ and $n = 0$. We note that the concentration within the cell at time t is n/V. We substitute into Eq. (A3.3) and rearrange to obtain

$$\ln[S_I/(S_I - n/V)] = PAt/V \qquad (A3.4)$$

At the half-time of entry, $t_{1/2}$, $n/V = S_I/2$. Substituting these values into Eq. (A3.4), we obtain

$$\ln 2 = PAt_{1/2}/V \tag{A3.5}$$

or

$$t_{1/2} = 0.693V/PA \tag{A3.6}$$

which is Eq. (2.12).

Index

Numbers in italics indicate data in tables.

317